what are they

lc

lj

Haven't done them
yet

STRENGTH
OF
MATERIALS
FOR
TECHNOLOGY

STRENGTH
OF
MATERIALS
FOR
TECHNOLOGY

CHARLES D. BRUCH, P.E.

Professor, Engineering and General Technology
Harrisburg Area Community College

JOHN WILEY & SONS, INC.
New York · Santa Barbara · Chichester · Brisbane · Toronto

Library of Congress Cataloging in Publication Data:

Bruch, Charles D
  Strength of materials for technology.

  Includes indexes.
   1. Strength of materials. 2. Strains and
stresses. I. Title.

TA405.B76   620.1'12   77-27629
ISBN 0-471-11372-7

To my esteemed friend and associate

Albert J. Magnotta
1910–1974

The momentum of his teaching continues to challenge
and inspire teachers and students.

PREFACE

This book is written for the student in engineering technology, the student in trade schools, and the student in postsecondary courses in vocational high schools.

It is assumed that those using this text have attained substantial fluency in algebra and trigonometry and has had one semester in mechanics. Although some students may have had some calculus and others may be studying it concurrently with this course, no knowledge of it is required for the use of this book.

Recognizing that technology students are by nature doers and likely to be interested in the immediate goal, long involved discourses are avoided. Principles and techniques are introduced for the most part by fully explained examples. Graphical techniques are presented that are helpful in solving problems without the aid of calculus. Examples are the extensive use of shear and bending moment diagrams in the determination of stresses and deflections in bending. Mohr's circle diagram is used in the determination of stresses due to the combination of tensile, compressive, bending, and torsional loading.

Answers to even-numbered problems are included at the end of the book.

Under the effect of forces, members of machines and structures compress, bend, buckle, flow, break, split, or crush. They may even become weaker or stronger. Effective technician-designers must have a "feeling" for the materials used in their designs. The purpose of the first chapter is to provide a brief introduction to a true understanding of materials. I believe that it is a necessary prelude to the study of stresses, strains, and deformations of materials that are caused by the various combinations of the forces to which they may be exposed.

The second chapter sets forth some of the basic concepts and laws relating to stress and strain and provides a bridge between the first chapter and the studies that follow. It concentrates on those areas of knowledge that the technician will find useful. Although steel is the most generally used load-bearing material, the importance and application of a number of materials are shown in the examples and problems. Too often, students are allowed to believe that if it isn't steel, it doesn't exist.

The world is going metric. In the near future engineering graduates will be metric-system oriented. Thus technicians who will be working with those engineers need to understand the metric system. This book is designed to be used for courses conducted entirely in the customary system, entirely in the metric system, or in any mix of the two. Appendix 1 at the end of the book describes the International Metric System–SI and provides conversion tables.

To aid students in knowing where they are going and in evaluating their progress, there are lists of objectives at the beginning of the chapters and accomplishment checklists at the ends of the chapters.

Charles D. Bruch

CONTENTS

CONTENTS

6. TORSION

7. SHEAR FORCE AND BENDING MOMENT IN BEAMS

CONTENTS

SYMBOLS USED IN THIS TEXT ————————————————

The symbols used in this text are in accordance with current industry usage as in The United States Steel Corporation's *Steel Design Manual* 1974; The Aluminum Association's *Aluminum Construction Manual*, 1972; and the American Plywood Association's *Plywood Design Specification*, 1974.

A cross-sectional area

A' area of cross section lying beyond shear plane with respect to the neutral axis

A_{cp} area of cut plane

A_n net area

A_s area in shear; also area of cross section

C_c column slenderness ratio dividing elastic and inelastic buckling

C_t coefficient of thermal expansion

D diameter

D_b bolt diameter

D_m mean diameter

D_h hole diameter

E modulus of elasticity

E_R modulus of resiliency

F force

F_a allowable axial stress

F_{cb} allowable bearing stress for bolt

F_{cf} allowable bearing stress for flange

F_v allowable shear stress

F_{vb} allowable shear stress for bolt

F_{vf} allowable shear stress for flange

FS	factor of safety
G	shear modulus of elasticity
hp	horsepower
I	moment of inertia
J	polar moment of inertia
K	effective length factor for column
L	length
M	bending moment; also turning moment
N	number of connectors, i.e., bolts, rivets; also number of revolutions per minute
N_s	number of holes in cross section of connection
P	load or force
P_a	allowable load
P_{bc}	force acting at bolt circle radius
P_c	critical load
P_t	tensile load
P_v	shear load
R	radius; also reaction of support
R_{bc}	radius of bolt circle
S	section modulus
T	torque
T_a	resisting torque on infinitesimal area
U	work
U_{st}	strain energy
U_{ust}	maximum strain energy (at $f = f_e$)
V	shearing force
W	weight or total load
b	section width

c distance from neutral plane to extreme fiber of beam

dA infinitesimally small area

$d_{B/A}$ vertical distance between deflected point B of a beam and the tangent to the curve of the beam at another point A.

e eccentricity

f stress—for the following stress symbols, f is replaced by F to indicate allowable stresses

f_a axial stress

f_b bending stress

f_c compressive stress; also buckling stress for columns

f_{ci} stress in circumferential section of cylinder.

f_{cc} cover plate bearing stress

f_{cm} main plate bearing stress

f_e stress at elastic limit

f_L stress in longitudinal section of cylinder

f_n tensile stress normal to given surface within object

f_p stress on particle

f_s maximum stress in thin-walled spherical pressure vessel

f_t tensile stress on section

f_u tensile strength; also average stress at ultimate load

f_{us} ultimate shear stress

f_v shear stress

f_{va} shear stress on infinitesimal area

f_{vb} shear stress in fasteners perpendicular to axis

f_{vf} shear stress in flange at surface of hub

f_{vxy} the unit shear stress on the x-plane and y-plane

f_y yield strength

f_x tensile or compressive stress induced by a force acting in the direction of the x axis

f_y tensile or compressive stress induced by a force acting in the direction of the y axis

h section height

n number of connectors, i.e., bolts, rivets

p pressure

q weight per unit of length

r minor radius; also radius of gyration

t thickness

t_w web thickness of beam

u Poisson's ratio

$\bar{x}$ distance from y axis to centroid or center of gravity in the direction of the x axis

$\bar{y}$ distance from x axis to centroid or center of gravity in the direction of the y axis

Δ elongation or deflection

ΔT change in temperature

α angular acceleration

ε strain

ε_a axial strain

ε_ℓ longitudinal strain

ρ radius of curvature

θ 1. angle of twist

        2. angle of tangent to the curve of a beam at a given point and the horizontal

γ angular deformation in radians

ω angular velocity

CHAPTER
OBJECTIVES ————————————————

The object of your study of this chapter is to learn

1. What is meant by the strength of a material.

2. The two types of failure of materials.

3. The meaning of the terms *elasticity*, *elastic limit*, and *permanent set*.

4. The difference between plasticity, ductility, and malleability.

5. The difference between a material simply described as stiff and one described as brittle.

6. What a metal is and the nature of its physical composition.

7. What alloys are and why they are used.

8. How metals and alloys can be changed in physical properties by heating and working.

9. The important advantages and disadvantages of using wood as a structural material.

10. The major advantages and disadvantages of using plastics as an engineering material.

11. The basic classifications of building stones.

12. Some of the significant engineering properties of granite, limestone, sandstone, marble, and slate.

13. The major types of brick and their uses.

14. What concrete is and the significance of the water-cement ratio.

CHAPTER——————————1
NATURE OF MATERIALS

Under the effect of forces, members of machines and structures compress, stretch, bend, buckle, flow, break, split, or crush. They may even become weaker or stronger. The effective technician-designer must have a "feeling" for the materials used in manufacturing and construction. By "feeling" I am referring to the understanding the designer develops concerning materials. This understanding is not just of their external measurable qualities but also of their inner natures, even their relationship to others in the society of materials. The purpose of this chapter is to provide an introduction, small as it may be, to the true understanding of materials.

1-1 DEFINITIONS OF PROPERTIES

strength

The *strength* of a material is its ability to withstand externally applied forces without failing. Failure of the material can be of two types: *catastrophic failure* such as breaking, crushing, or buckling or *gradual failure* deforming to a point where it cannot serve the purpose for which it was intended. An example of the latter is the failure of a spring which has stretched beyond usefulness by being repeatedly overloaded.

NATURE OF MATERIALS

elasticity

When a load is placed on an elastic member of a structure or machine the member changes in size or shape. When the load is removed the member may return to its original size and shape. *Elasticity* is the ability of a material to return to its original size and shape after having all loads removed from it. When you drive your car across a bridge, the bridge sags under the car's weight. When your car leaves the bridge, the bridge springs back to its original shape due to the elasticity of its structural members.

When an elastic body is loaded beyond a certain point it will not return to its original size and shape. The point beyond which a material will not return to its original dimensions is its *elastic limit*. When a material is deformed beyond its elastic limit the difference between its original dimensions and its final dimensions—each at zero load—is called its *permanent set*.

plasticity

Plasticity is the opposite of elasticity. A plastic material when deformed will not spring back when the load causing the deformation is removed. Putty is the prime example. Lead is a common plastic engineering material. There probably are no perfectly plastic engineering materials; even the foregoing examples exhibit some though insignificant elasticity. Elastic materials, when deformed beyond their elastic limits, undergo *plastic deformation*. The result of plastic deformation is permanent set.

ductility

A material which will undergo considerable deformation without breaking is said to be *ductile*. The steel bodies of automobiles are formed by stamping sheets of steel between dies to give them the desired shapes of fenders, roofs, doors, and so on. During this shaping the steel is greatly stretched without tearing. To do this the steel must be ductile. Thus we see that ductility is a very important property. It is of interest that while the steel is being plastically formed, it increases in strength and loses ductility.

malleability

Malleability is the property that allows a material to be formed into different shapes by hammering or otherwise compressing it. Most ductile materials are malleable to some degree. On the other hand, a considerable number of malleable materials have no significant ductility.

stiffness

Stiffness is that quality of a material which enables it to resist deformation. Under the same load steel will deform less than wood because steel is stiffer than wood. Tool steel will deform less than structural steel under the same load because it has greater stiffness than structural steel. For service where any significant change in shape would result in the failure of a material to do its job, stiffness would be of the utmost importance. *Brittle* materials are stiff materials which cannot absorb much of a load, particularly a shock load, without breaking.

other properties

There are numerous other properties that may be important for different applications. Some of these are; corrosion resistance, heat resistance, wear resistance, fatigue resistance, weight, melting point, flame resistance, thermal expansion and contraction, electrical conductivity, resilience, color, porosity, and weldability. Some of these will be considered and defined later in this book.

The accomplished designer develops a broad knowledge of large variety of materials. A very significant statement is made in a publication of the nation's largest steel company. "The best guide in selecting a steel that is appropriate for a given application is experience with existing and past structures."[1] The same statement can reasonably be applied to materials in general. This is not to be interpreted that new and even untried materials should not be used, but that their selection should be made in relation to existing and tried materials of similar properties. Sound progress in materials application is usually gradual. The astute designer develops a sixth sense concerning materials. To

[1] USS Steel Design Manual, R. L. Brockenbaugh and B. G. Johnston, United states Steel Corporation, 1974, p. 27.

develop this one must absorb as much knowledge as possible concerning the properties and use of materials.

1-2 METALS

Although there are over 60 metallic elements in nature there are but a handful that are both physically and economically suitable for commercial use in their elemental form. They are almost always used in combination with other metals and nonmetals. These combinations are called *alloys*. The most commonly used metal—steel—is an alloy of iron, carbon, and a number of other elements.

Metals are composed of crystals, the sizes of which determine some of the principle properties of the metals such as hardness. By the process called heat-treating, during which the metals are alternately heated and cooled, the properties of the steel, including crystal size (grain size), can be changed to suit the particular application. Even the crystal structure of pure metals is not perfect. Surprisingly, this is desirable. For example, metals could not be deformed plastically if stray atoms did not exist amid the crystal structure. Aside from the crystal imperfections which are desirable there are inclusions of foreign matter which are undesirable. Great pains and considerable expense are incurred in eliminating as much of these undesirable impurities as is economically feasible.

Metals are formed into usable shapes by processes such as casting, rolling, forging, sintering, drawing, flame cutting, stamping, and machining. These processes may cause the physical characteristics to change in a number of ways. Forming operations carried out under "cold" conditions (cold forming) increase the tensile strength and hardness. Rolling processes elongate the crystals in the direction of rolling, thus causing the fatigue resistance to be greater in the direction of rolling than it is in the direction perpendicular to rolling. In forging the metal flows and the crystals elongate in the directions of flow; intelligent forging design can take advantage of this to attain highly desirable strength characteristics. Powder metallurgy parts are produced by a three-step process in which elemental metal powders or alloys are blended together, are compacted under pressure to the desired shape, and are then sintered (heated in a protective atmosphere furnace to a relatively high temperature, but below the melting point of the metal). In the powder metallurgy process the density and porosity of the metal can be controlled. Porous bearing materials and sound-reducing gears are examples of powder metallurgy products.

1-3 WOODS

Wood is the material with which we are most familiar. It is a cellular fibrous material which, in its natural state, comes in a wide range of strengths and hardnesses. Our forefathers used logs to build houses, many of which are still standing today. For our use today wood is cut, rough sawed, dried, and then sized and finished. Its properties parallel to the grain vary greatly from those perpendicular to the grain. It is an excellent structural material because it has excellent qualities of strength and rigidity combined with comparative lightness of weight. It can be laminated with powerful adhesives to make frames for tennis rackets or plywood for boats, buildings, and any number of useful products. A most interesting application of laminated structural members is the prefabrication of large integral columns and beams used in many modern churches and recreational buildings. These members are both attractive and, being very strong and fire resistant, are structurally desirable. An irradiated wood-plastic material called novawood is several times as hard as untreated wood and can be formed and finished by the same techniques used on ordinary wood. It is made by impregnating ordinary wood with a monomer (the basic building block of plastics) and irradiating it with gamma rays to link the monomers into chains, thus forming a strong impervious material.

1-4 PLASTICS

Plastics abound everywhere—there seems to be no end of them. They have even come to be an environmental problem. Nevertheless, plastics have been an immeasurable boon to humanity and will continue to be increasingly valuable. Our interest lies in the *engineering plastics*. Engineering plastics have been defined as those which possess physical properties enabling them to perform for a long time in structural applications, over a wide temperature range, under mechanical stress, and in difficult chemical and physical environments.

Plastics have good chemical and moisture resistance. However, some of them will absorb moisture or solvents and when they do, they may change in dimension. This can be a disadvantage for parts such as gears which must maintain their shape within very close limits. Plastics have good vibration and shock resistance and have greater wear and abrasion resistance than some metals. A big plus for plastics is the ease with which they can be fabricated, which frequently makes them less costly than equivalent parts made of metals. Among their significant disadvantages are their relatively high thermal expansions, their brittleness under cold conditions, and the degradation of most of them by sunlight. To overcome some of the natural deficiencies of plastics other substances called fillers are added to their mix to form products with greatly improved properties such as impact strength and dimensional stability. Some of the fillers used in plastics are glass fibers, asbestos, carbon black, metal powders and fibers, talc, wood flour, and fabrics.

Plastics find their place in nearly all areas of engineering design. No other class of engineering material has as broad a range of properties and applications as plastics. For their low friction characteristics some plastics are used for such things as bearings and the slides of high-speed elevators. Other plastics having high tensile strengths and resistance to fatigue are used for gears and components of automobile door latches. Electrical connectors such as the plugs and sockets in house wiring have long been made of plastics. Also well known are the plastics known commercially as lucite and plexiglass which are crystal clear and are used extensively for aircraft and marine windows.

1-5 STONE, BRICK, AND CONCRETE

Stone, brick, and concrete are called *inert materials* because they do not react chemically with the environment as do the majority of structural materials. The above heading lists these materials in the order of increasing importance.

stone

Building stones are classified to their origin—*igneous, sedimentary,* and *metamorphic.* Of the *igneous* rocks, granite is the important one. *Granite* is composed mostly of hard minerals strongly welded together by the great forces of the volcanic action that formed them. It is the strongest stone in common use and is extremely resistant to weathering. The most widely distributed and used stone is limestone. *Limestone* is a sedimentary stone that appears in a wide variety of colors and compositions. Its physical qualities vary quite widely and some limestones are not suitable for structural use. *Sandstones,* like limestones, vary greatly in color, texture, durability, and usefulness. The metamorphic rocks, *marble* and *slate,* are not used as load-bearing structural members. Marble is used extensively for decoration. Slate is occasionally used for roofing and it is still used for blackboards.

brick

Bricks, in addition to their decorative uses and structural uses in buildings, are used industrially as linings for high-temperature vessels and chemical containers of many varieties. Obviously these varied uses require bricks of widely differing qualities. *Structural brick* is of two compositions, *clay brick* and *sand-lime brick*. Structural bricks vary in strength and durability according to the type of materials used, the method of molding, and the method of firing or curing. In general the compressive strengths of clay and sand-lime bricks are equivalent for equivalent grades. *Refractory bricks* are made to withstand temperatures up to 1000°C while maintaining good load-bearing strength. It is also important for refractory bricks to have good insulating qualities so that they will aid in containing the heat in the vessels in which they are used. Chemical process industries use acid-resisting bricks in quite large quantities. Many refractory materials are used in gas turbines, jet engines, and rockets where heats are so high that metal parts will not withstand them.

concrete

Concrete is a mixture of cement, sand, gravel, and water, in which the water and cement combine in a chemical reaction to form a very hard and adhesive material which fills in the open spaces between the particles of sand and gravel and binds them together in a mass much like stone. Like stone and brick, concrete is strong in compression but comparatively weak in tension. The two most important qualities of concrete are its strength and durability. The ratio of the amount of water to the amount of cement, the *water-cement ratio*, is the principal controlling factor determining the strength and durability of concrete. The final determination of the amount of sand and gravel to be used is largely a cut and try process.

In most applications concrete is reinforced with metal embedded in it in such a manner that they act together. The concrete supports the compressive load, while the steel withstands the tensile load. This combination of concrete and steel is called *reinforced concrete*. The reinforcing steel may be in the form of expanded metal, wire mesh, or bars. Reinforcing bars have a pattern of ribbing on their surfaces to aid the bonding of the steel and concrete.

When a load is placed on top of a beam, which is the usual case, the bending of the beam causes the top of the beam to be in compression while the bottom of the beam is placed in tension. If the load stretches the steel reinforcing to the extent that the concrete is placed in tension, the concrete may crack. To enable reinforced concrete to carry greater loads a technique has been developed whereby concrete beams are precast with the reinforcing rods being held in tension until the concrete hardens around them. Thus the beams are initially compressed. This compression must be relieved before any tension is placed on the concrete. Beams manufactured by this process are called *prestressed concrete beams*.

1-6 SELECTION OF MATERIALS

"What material shall I use?" and "Is there a better material for the job?" are constantly recurring questions in the minds of both technicians and engineers. At the outset it might appear that there should be pat answers to these questions. Many times there are, but in many cases there are not. In addition to the physical abilities of the material to do the job, there are numerous other considerations such as esthetics, economics, and availability. In many cases the selection will be a compromise. The most desirable material may be too costly or it may not be available when needed.

In most cases economics is an important consideration—in many cases it is the most important consideration; in all cases it should be seriously considered. Many factors enter into the cost of a material. First, the price per pound, ton, or kilogram at the supplier's mill, factory, or warehouse; the cost of transportation from the supplier's place of business to the plant or construction site; cost of any special protection necessary during transportation; cost of carrying it in inventory before actual use; cost of preparing it for use; cost of machining and fabrication; cost of maintaining it during its useful life; cost of replacement due to deterioration or obsolescence. The list could go on and on.

An interesting example of the economics of material selection arises as the result of high-strength steels. These steels find economical use in places where the steel is undergoing tensile or bending loads. For example, North Carolina saved about $13 per lineal foot on the bridge spanning the Roanoke river between Oak City and Lewiston by using high-strength steel instead of structural carbon steel. The designer must consider "cost per unit of strength" and "weight per unit of strength." The importance in considering weight is that in a large structure a significant portion of its strength is required to support its own weight. Thus, a saving in weight reduces the amount of steel or other supporting material needed and results in a cost saving.

Bridges made of ordinary steel will rust if the steel is not protected by a rust-preventive coating. With the exception of maintaining the roadbed the most costly maintenance is that of keeping the bridge painted. Weathering steels are now available that form their own natural protective coating and do not require painting. When the high cost of periodically painting structures having exposed steel is considered, the use of weathering steels becomes economically attractive.

Wherever weight is a disadvantage the use of lightweight materials is likely to prevail. This use of lightweight materials in the transportation industry is likely to increase considerably as the cost of fuels increases. The bodies of trucks and buses are largely made of aluminum. In aircraft, where the requirements are more critical, aluminum, magnesium, titanium, and plastics are extensively used. In all cases the use of these lightweight materials is a matter of economics. Every pound saved in the construction of the body of a transportation vehicle permits one pound more payload or at least reduces by a pound the deadload consuming fuel. If you are designing transportation equipment, lightweight is not the only consideration in selecting the proper materials. You must also consider such things as durability, strength, fatigue resistance, formability, machinability, sound transmission, heat transmission, weldability, resistance to chemicals, and availability in the quantities and properties desired, as well as appearance.

1-7 ACCOMPLISHMENT CHECKLIST

On the basis of your studies of this chapter do you know:

1. What is meant by the "strength of a material"?

2. The two types of failure of materials?

3. The meanings of the terms *elasticity, elastic limit,* and *permanent set*?

4. The difference between plasticity, ductility, and malleability?

5. The difference between a material simply described as stiff and one described as brittle?

6. What a metal is and its physical composition?

7. What an alloy is and why they are used?

8. How metals and alloys can be changed in physical properties by heating and working?

9. The important advantages and disadvantages of using wood as a structural material?

10. The major advantages and disadvantages of plastics as an engineering material?

11.   The basic classifications of building stones?

12.   Some of the significant engineering properties of granite, limestone, sandstone, marble, and slate?

13.   The major types of brick and their uses?

14.   What concrete is and the significance of the water-cement ratio?

1-8 SUMMARY

The *strength* of a material is its ability to withstand externally applied forces without failing. Failure can be *catastrophic* such as breaking, crushing, or buckling, or *gradual* such as by being deformed to the point of uselessness by repeated overloading.

Elasticity is the ability of a material to return to its original size and shape after having been deformed. The point beyond which a material will not return to its original dimensions is its *elastic limit*. When a material is loaded beyond its elastic limit the difference between its original and final dimensions, each at zero load, is called its *permanent set*.

Plasticity is the opposite of elasticity. Elastic materials, when deformed beyond their elastic limits, undergo *plastic deformation*. *Ductility* is the property that enables a material to undergo considerable deformation without breaking. *Malleability* is the property that allows a material to be formed into different shapes by hammering or otherwise compressing it. *Stiffness* is the quality of a material that enables it to resist deformation. *Brittle* materials cannot absorb much of a shock load without breaking.

Metals are elements made up of crystals. Combinations of metals with other metals or nonmetals are called *alloys*. Engineering alloys generally have properties superior to any of the ingredients from which they are compounded. Forming operations can change the characteristics of metals and alloys in many ways such as increasing their strength and hardness or changing their grain structure.

Wood is an excellent structural material which has significantly different properties in the direction of the grain when compared with those in the direction across the grain. It is frequently used as beams in structures when it is laminated, laminated beams being strong, light in weight, and fire resistant.

Engineering plastics are those which possess physical properties enabling them to perform for a long time, over a wide range of temperatures, under mechanical stress, and in difficult chemical and physical environments. No other class of engineering materials has as broad a range of properties and applications as do plastics.

Stone, brick, and concrete are called *inert materials*. Building stones are classified according to their origin—*igneous*, *sedimentary*, and *metamorphic*. *Granite* is composed mostly of hard materials volcanically welded together. *Limestone* is a sedimentary stone having widely varying physical qualities. *Marble* and *slate* are metamorphic rocks and are not suitable for load-bearing. Bricks are used in structures, as linings for high-temperature vessels, and as acid-resisting linings in the chemical process industries. *Structural brick* is of two compositions, *clay brick* and *sand-lime brick. Refractory bricks* are made to withstand temperature up to 1000°C while maintaining good load bearing strength.

Concrete is a mixture of cement, sand, gravel, and water. The ratio of the amount of water to the amount of cement, the *water-cement ratio*, is the principal controlling factor determining the strength and durability of concrete. *Reinforcing bars* are used to overcome the tensile weakness of concrete. When concrete beams are constructed so as to place the beams initially in compression (which must be relieved before any tension is placed on the concrete), the beams are called *prestressed concrete beams.*

In the selection of materials such things as durability, strength, fatigue resistance, formability, machinability, sound transmission, heat transmission, weldability, resistance to chemicals, and availability must be considered. Most important is economics. A designer must consider such things as "cost per unit of strength" and "weight per unit of strength."

CHAPTER
OBJECTIVES

The purpose of your study of this chapter is to

1. Understand the natures of stress and strain.

2. Understand the relationship between stress and strain.

3. Learn Hooke's law and how to apply it to practical problems.

4. Learn the significance of Poisson's ratio and how to use it to determine deformations of bodies along axes perpendicular to the direction of the applied forces.

5. Learn how most materials expand and contract due to changes in temperature and how to determine the forces created in constrained bodies due to these changes.

CHAPTER ——————————— 2
STRESS AND STRAIN

In the study of mechanics the effect of forces on bodies at rest and in motion was studied in a quite limited fashion. All bodies were considered to be perfectly rigid. We determined the forces acting on, in, and through members of structures and machines. We did not consider what changes these forces caused in the shape of the members. These changes are the very essence of the study called Strength of Materials. The two most significant terms used in Strength of Materials are stress and strain.

2-1 NATURE OF STRESS

Figure 2-1 shows a beam which is a member of a truss, the adjacent portion of which is shown by the phantom lines. Our problem is to determine what is going on inside the member. Now, the exact internal structure of the member is much too complex to deal with and actually varies from one point to another. We therefore assume it to be a homogeneous mass of molecules, each molecule being held in position by cohesive forces. The force being exerted on this beam by the other members of the structure is tending either to push its molecules together or pull them apart. Due to the cohesive forces each molecule will resist any change in its position. Considering the plane Y, which cuts the member perpendicular to its axis, the resistance of the molecules to separating on that plane must in total equal the force acting through the member at that point, otherwise the beam would collapse or separate. This resistance to change by the molecules is called *stress*. We

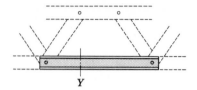

FIGURE 2-1

evaluate stress by determining the resistance of the molecules over one square unit of the plane Y. Thus the stress is

$$f = \frac{P}{A} \qquad (2\text{-}1)$$

where f is the stress, P is the force perpendicular to and acting through the centroid of the plane of the stress, and A is the area of the cross section over which the stress is being determined.

In the determination of stresses and deformations of elastic bodies we are not at liberty to move forces about as is permissible by the laws of the mechanics of rigid bodies. Changing the points of application of forces on elastic systems can cause significant changes in stresses and deformations.

When a beam is in tension (being pulled on) we say that the stress is *tensile stress*. When a member is under compression (being pushed on) we have *compressive stress*, and when the force acts parallel to the plane of the stress it is *shearing stress*.

Example

customary

Determine the stress in the beam in Figure 2-1 if the beam is a W10X100 and is acted on by a tensile force of 58,000 lb.

metric

Determine the stress in the beam in Figure 2-1 if the beam is a 300 × 200 – 65.4 wide-flange beam and is acted on by a tensile force of 260 kN.

Solution

customary

The W in the beam designation indicates that it is a wide flange beam (see Appendix 2). The 10 reveals that its nominal depth, measured parallel to the web, is

metric

The 300 in the beam designation indicates that it has a nominal depth of 300 mm. The 200 reveals that its nominal flange width is 200 mm, approximately 10 in., the exact depth being given in Appendix 2. The 100 is the beam's weight in pounds per foot of length and identifies the beam as a nominal 10×10.

Since the weight of the beam is so small in comparison with the 58,000 lb force, it would not show up in the answer and may therefore be ignored.

We find the cross-sectional area of our beam as given in Appendix 2 to be 29.4 in.^{2} So the stress in the beam is

$$f = \frac{P}{A} = \frac{58,800 \text{ lb}}{29.4 \text{ in.}^2} = 2000 \text{ psi}$$

where psi is *pounds per square inch.*

whereas the 65.4 is the beam's weight in kilograms per meter of length. The exact dimensions are given in Appendix 2.

Since the weight of the beam is so small in comparison with the 260 kN force it would not show up in the answer and may therefore be ignored.

We find the cross-sectional area of our beam as given in Appendix 2 to be 83.36 cm^2. So the stress in the beam is

$$f = \frac{P}{A} = \frac{260 \text{ kN}}{83.36 \times 10^{-4} m^2} = 31.2 \text{ MPa}$$

where MPa (megapascals) is meganewtons per square meter.

In the foregoing examples the beam was loaded so that the force was evenly distributed over its cross section. Even with axial loading such as this the actual distribution of the load is not perfectly uniform. Loads may be concentrated on a small portion of the cross section at the point of application. Nevertheless, it is assumed that at a comparatively short distance from the point of application the load does become evenly distributed. The error incurred by this assumption should be insignificant.

Members of machines and structures are usually loaded in ways that cause significant differences in the stresses from one part of a cross section to another. An example is the bending of a beam. At the top of the beam the stress is compressive, whereas at the bottom it is tensile, with some point in between having zero stress. The method of dealing with this will be taken up in later chapters.

2-2 NATURE OF STRAIN

When materials are subjected to forces they deform. *Strain* is the deformation per unit of length of a member in any specified direction due to the load on that member. The number of molecules in the beam in the foregoing example did not change, therefore the distance between the molecules must have changed. Again we have no way of determining the microscopic changes in the distances between the molecules but we can determine the total change

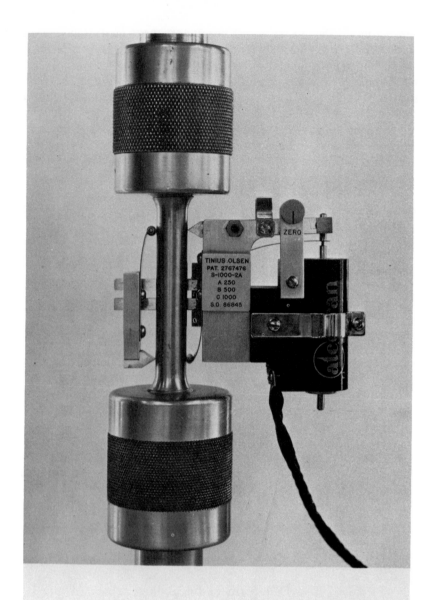

Typical Model S-1000 averaging type extensometer
shown in place on standard specimen. Tinius Olsen
Testing Machine Co., Inc., Willow Grove, Pa. TO4664

FIGURE 2-2 (Photo courtesy Tinius Olsen Testing Machine Co., Inc.)

over a unit of length such as the inch or meter. The change in dimension can be determined by mechanical or electronic devices. One of these devices, the extensometer, is shown in Figure 2-2. This device measures the change in length of a test sample over a given length. The strain on actual members of structures and machines is usually measured by the use of electronic strain gages such as are shown in Figure 2-3. The strain gage is glued to the surface of the part being tested. As the load is applied and the part deforms the strain gage deforms with it. As the strain gage deforms its electrical resistance changes and this change in resistance is electronically translated to report the strain being incurred by the part due to the loading. On large members where it is important to determine the strain in different locations on the part, a number of strain gages may be operated simultaneously to record this information. The mathematical relationship between total deformation and strain is

$$\epsilon = \frac{\Delta}{L} \qquad (2\text{-}2)$$

where ϵ is the strain, Δ is the total deformation, and L is the length in the direction of deformation.

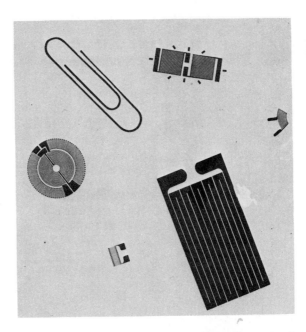

FIGURE 2-3 (Photo courtesy of Micro-Strain, Inc.)

2-3 RELATIONSHIP OF STRESS AND STRAIN

As the stress in a member increases the strain increases. In some instances the strain is proportional to the stress; at other times it is not. The relationship of stress and strain is best revealed by the stress-strain diagram. Figure 2-4 shows such a diagram for a typical elastic material. This diagram is developed by taking a precisely machined specimen, Figure 2-5 of the material and

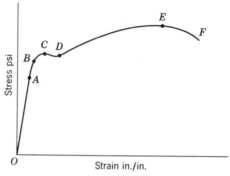

FIGURE 2-4

placing increasing tensile loads on it. The specimen, called a tensile test specimen, is usually about 9 in. long in customary units and has a precisely uniform diameter of 0.505 in. except at its ends which are of larger diameter and are threaded for mounting in the tensile test machine which places increasing loads on it until it fails. For every increment of load imposed on the specimen its strain is measured by an extensometer and the stress is calculated. The stress-strain curve is the plot of these values, stress being plotted on the vertical axis and strain on the horizontal.

Example

customary

If the unloaded length of the beam in the previous example were 10 ft and it was stretched 0.008 in. by the 58,800 lb load, the strain would be

$$\epsilon = \frac{\Delta}{L} = \frac{0.008 \text{ in.}}{120 \text{ in.}} = 0.000\,066\,7 \text{ in./in.}$$

metric

If the unloaded length of the beam in the previous example were 3 m and it was stretched 0.000 2 m by the 260,000 N load, the strain would be

$$\epsilon = \frac{\Delta}{L} = \frac{0.000\,2}{3\,m} = 0.000\,066\,7 \text{ m/m}$$

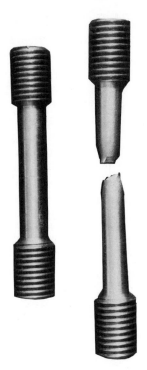

FIGURE 2-5   Steel test specimen before and after pulling.

Between the origin of the curve and point *A* the curve is straight showing that the strain is directly proportional to the stress. Point *A* is the *proportional limit* of the material. Between points *A* and *B* the curve is not straight revealing that beyond point *B* strain is no longer directly proportional to stress. Therefore, point *B* is called the *elastic limit*. Beyond point *B* the material deforms both plastically and elastically. If the load is released, the speciman will not return to its original dimension but will return to a point short of that. The difference between the original and new lengths at zero load is called the *permanent set*. When the stress reaches point *C* the speciman will continue to deform without any further application of load. In some cases there will be a slight dip in the curve between points *C* and *D*. This means that the material is deforming at a lower load than that at point *C*. This continued deformation without an increase in load is called *yielding* and point *C*, where the yielding begins, is the *yield point*. During the period of yielding there is sliding between the crystals of the material until finally they come to a point where the material can again resist extension. This happens at point *D*. Beyond *D* the curve regains its upward movement to point *E*, the highest point of the curve. The stress at point *E* is the *tensile strength*, also called the *ultimate strength* of the material.

It is important to know that the values of stress on the stress-strain curve beyond the elastic limit are not true values. When the material passes the elastic limit a significant reduction in the diameter of the central portion of the specimen begins. This is called "necking down" and is evident in the photograph of the fractured specimen in Figure 2-5. This necking down continues until the specimen is finally fractured. However, all of the stress calculations are made using the original cross section. If the curve were to be drawn by taking the necking down into consideration in the calculation of the stress, the curve would appear as in Figure 2-6. After passing through the yielding phase the material actually increases in strength. Point *F* is where the specimen ruptures, the stress at that point being called the *rupture strength*.

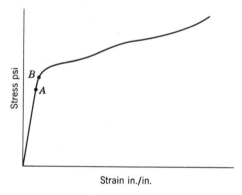

Strain in./in.

FIGURE 2-6

Listed in the tables of physical properties available from the producers of materials is the *percentage elongation* which is usually stated as the *percentage elongation in 2 in. or 50 mm*. Standards for the tensile test usually specify that the elongation of the central 2 in. or 50 mm portion of the test specimen be measured. The percentage elongation is equal to the total elongation of the gage length at the time of rupture divided by the original gage length times 100. For example, if the total elongation of a 2017-T4 aluminum alloy was 0.22 in. at the time of rupture, the percentage elongation would be 0.22 in./2 in. × 100 = 11 percent.

Most materials do not have as distinct a yield point as that of Figure 2-5. Therefore, an arbitrary point is selected; this point being determined by the amount of permanent set that is acceptable in design work. The amount usually accepted for steel is 0.2 percent. The arbitrary yield point is then determined by drawing a straight line from the 0.002 or 50 μm point on the strain axis of the stress-strain diagram parallel to the stress-strain curve at its origin. The point at which this line intersects the stress-strain curve is the yield point at 0.2% offset. This method is called the *offset method* of determining the yield point.

2-4 HOOKE'S LAW

The straight-line proportionality of stress to strain was first observed by an English scientist, Robert Hooke, about 1658. Approximately 150 years later another English scientist, Thomas Young, determined the constant of this proportionality, thus enabling us to state it in the form of the equation

$$f = E\epsilon \tag{2-3}$$

where f is the unit stress, ϵ the unit strain, and E is the *modulus of elasticity*.

The above equation in the form given has limited use because we are most often interested in the total deformation of a member due to having some load placed on it. To put this equation in a more generally usable form we determine the values of unit stress and unit strain and express them in terms of the load and total deformation of the member.

$$f = \frac{P}{A} \quad \text{and} \quad \epsilon = \frac{\Delta}{L}$$

where P is the total load on the member, A is the cross-sectional area of the member, Δ is the total elongation of the member, and L is the length of the member. Substituting these values in Equation (2-3) gives us

$$\frac{P}{A} = E\frac{\Delta}{L}$$

Solving this for Δ, which is what we usually want to know, it becomes

$$\Delta = \frac{PL}{AE} \tag{2-4}$$

or, since $P/A = f$,

$$\Delta = \frac{fL}{E} \tag{2-4a}$$

Example

customary

A 10-ft-long wide flange W12X92 column supports a 510 kip axial load in the interior of a single story building. Determine the total deformation of the column if it is made of A36 steel.

metric

A 3-m-long rolled wide flange W30X135 column supports a 2.30 MN axial load in the interior of a single story building. Determine the total deformation of the column if it is made of A36 steel.

Solution

customary

From Appendix 3, $E = 29,000,000$ psi
From Appendix 2, $A = 27.1$ in.^{2}
Given: $P = 510,000$ lb and $L = 10$ ft $= 120$ in.
Substitute these values in Equation 2-4

$$\Delta = \frac{PL}{AE} = \frac{510,000 \text{ lb } (120 \text{ in.})}{27.1 \text{ in.}^2(29 \times 10^6 \text{ psi})} = 0.079 \text{ in.}$$

Another situation that may occur is that where the member does not have the same cross section throughout.

metric

*From Appendix 3, $E = 200 \times 10^9$ Pa
From Appendix 2, $A = 0.017\,5\,m^2$*

Then

$$\Delta = \frac{PL}{AE} = \frac{(2.30 \text{ MN})(3 \text{ m})}{(0.017\,5 \text{ m}^2)(200 \times 10^9 \text{ Pa})}$$
$$= 0.001\,97 \text{ m}$$

Another situation that may occur is that where the member does not have the same cross section throughout.

Example

customary

The continuous aluminum rod in Figure 2-7 is made of 2024-T4 aluminum. Determine the elongation caused by the 2000 lb load.

metric

The continuous aluminum rod in Figure 2-7 is made of 2024-T4 aluminum. Determine the elongation caused by the 900 kg load.

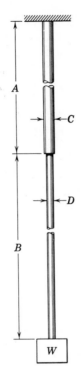

Dimensions		
	Customary	Metric
A	10 ft	3 m
B	20 ft	6 m
C	1" dia	25 mm dia
D	¾" dia	20 mm dia
W	2000 lb	900 kg

FIGURE 2-7

Solution

customary

The weight of the rod itself will be neglected.

The force throughout the rod is 2000 lb.

Designate the 10-ft-section by subscript 1 and the 20 ft section by subscript 2.

The total elongation of the rod is the sum of the elongations of its two sections.

$$\Delta = \frac{P_1 L_1}{A_1 E_1} + \frac{P_2 L_2}{A_2 E_2}$$

$P_1 = P_2 = 2000 \text{ lb} \quad A_1 = 0.785 \text{ in.}^2 \quad A_2 = 0.442 \text{ in.}^2$

metric

The weight of the rod itself may be neglected as insignificant.

The force throughout the rod is $900 \times 9.81 = 8\,830 \text{ N}$.

Designate the 3 m section by subscript 1 and the 6 m section by subscript 2.

The total elongation of the rod is the sum of the elongations of its two sections.

$$\Delta = \frac{P_1 L_1}{A_1 E_1} + \frac{P_2 L_2}{A_2 E_2}$$

STRESS AND STRAIN

From Appendix 3 we find E_1 and E_2 to be 10.6×10^6 psi.

$$L_1 = 120 \text{ in. } L_2 = 240 \text{ in.}$$

Substituting the above values in the equation gives us

$$\Delta = \frac{2000 \text{ lb } (120 \text{ in.})}{0.785 \ (10.6 \times 10^6)}$$

$$+ \frac{2000 \text{ lb } (240 \text{ in.})}{0.442 \text{ in. } (10.6 \times 10^6)} = 0.131 \text{ in.}$$

$P_1 = P_2 = 8\,830 \text{ N}, \quad A_1 = 0.000\,491 \text{ m}^2,$
$A_2 = 0.000\,314 \text{ m}^2$

From Appendix 3 we find E_1 and E_2 to be 69×10^9 Pa

$$L_1 = 3 \text{ m and } L_2 = 6 \text{ m}$$

Substituting the above values in the equation gives us

$$\Delta = \frac{8\,830 \text{ N } (3 \text{ m})}{0.000\,491 \text{ m}^2 \ (69 \times 10^9 \text{ Pa})}$$

$$+ \frac{8\,830 \text{ N } (6 \text{ m})}{0.000\,314 \text{ m}^2 \ (69 \times 10^9 \text{ Pa})}$$

$$\Delta = 0.003\,23 \text{ m}$$

2-5 POISSON'S RATIO

The application of a load to a body causes it to undergo strain in the direction of the applied load (axial strain), while it causes strain in the direction perpendicular to the applied load (lateral strain). This was discovered in 1828 by the French mathematician, Poisson. Within the elastic range of the material the ratio of the lateral strain to the axial strain remains constant. This ratio is called *Poisson's ratio*.

$$u = \frac{\epsilon_\ell}{\epsilon_a} \tag{2-5}$$

where u Poisson's ratio, ϵ_a is axial strain, and ϵ_ℓ is lateral strain.

Example

customary

In the example on page 000 it was found that a 510 kip axial load caused a W12X92 column 10 ft high to compress 0.075 in. Determine the change in the depth of this wide-flange member due to the load.

metric

In the example on page 000 it was found that an 8 830 N axial load caused a W30X135 column 3 m long to compress 0.001 88 m. Determine the change in the depth of this wide flange member due to the load.

28

Solution

customary

First we must determine the axial strain by use of Equation 2-2 page 000.

$$\epsilon_a = \frac{\Delta}{L} = \frac{0.075 \text{ in.}}{120 \text{ in.}} = 0.000\,625 \text{ in./in.}$$

Poisson's ratio for steel is approximately 0.3. Then from Equation 5

$$\epsilon_\ell = \epsilon_a u = 0.000\,625\,(0.3)$$
$$= 0.000\,187 \text{ in./in.}$$

From Appendix 2 we find the depth of the W12X92 column to be 12.62 in. Therefore, the change in the depth of the member is

$$\Delta = \epsilon L = 12.62 \text{ in. } (0.000\,187)$$
$$= 0.002\,37 \text{ in.}$$

metric

First we must determine the axial strain by use of Equation 2-2 page 000.

$$\epsilon_a = \frac{A}{L} = \frac{0.001\,88 \text{ m}}{3 \text{ m}} = 0.000\,627 \text{ m/m}$$

Poisson's ratio for steel is approximately 0.3. Then from Equation 5

$$\epsilon_\ell = \epsilon_a u = 0.000\,627\,(0.3)$$
$$= 0.000\,188 \text{ m/m}$$

From Appendix 2 we find the depth of the W30X136 column to be 32.00 cm. Therefore, the change in the depth of the member is

$$\Delta = \epsilon L = 32 \text{ cm } (0.000\,188) = 0.006\,02 \text{ m}$$

2-6   TEMPERATURE EFFECTS

Most engineering materials tend to expand when heated and retract when cooled. This must be considered in the design of structures and machines. If members are prevented from moving freely (are constrained), stresses will develop which, in some cases, might be intolerable and cause failures. The amount of strain in inches per inch of length per degree temperature change is called the *coefficient of thermal expansion* C_t. For an unconstrained member the change in length can be determined by

$$\Delta = C_t L(\Delta T) \tag{2-6}$$

where Δ is the total change in length, C_t is the coefficient of thermal expansion, L is the length of the member before the change in temperature, and (ΔT) is the change in temperature.

Example

customary

A steel plate-girder bridge has a span 400 ft long. How much change is there in its length from the coldest day in winter ($-20°F$) to the hottest day in summer ($92°F$)?

metric

A steel plate-girder bridge has a span 120 m long. How much change is there in its length from the coldest day in winter ($-30°C$) to the hottest day in summer ($35°C$)?

Solution

customary

From Appendix 4 we find that the coefficient of thermal expansion C_t for structural steel is 7.2×10^{-6} in./in./°F.

$$L = 400 \times 12 = 4800 \text{ in.}$$
$$\Delta T = 92° - (-20°) = 112°$$

Substituting these values in Equation 2-6 we get

$$\Delta = C_t L(\Delta T) = (7.2 \times 10^{-6} \text{ in./in./°F})$$
$$\times (4800 \text{ in.})(112°F) = 3.87 \text{ in.}$$

metric

From Appendix 4 we find that the coefficient of thermal expansion C_t for structural steel is 1.30×10^{-5} m/m/°C.

$$L = 120 \text{ m}, \ \Delta T = 35° - (-30°) = 65°C$$

Substituting these values in Equation 2-6 we get

$$\Delta = C_t L(\Delta T) = (1.30 \times 10^{-5} \text{ m/m/°C})$$
$$\times (120 \text{ m})(65°C) = 0.101 \text{ m}$$

If a body is not permitted to expand while incurring a change in temperature, the force that it would exert on whatever is constraining it would equal the force necessary to deform the body the amount that it would otherwise have expanded. Mathematically this can be stated

$$C_t L(\Delta T) = \frac{PL}{AE}$$

Because we are interested in the force, we will solve the above equation for P.

$$P = C_t(\Delta T)AE \tag{2-7}$$

Example

customary

The 4 in. diameter shaft of a gyrating flour sifting machine was held in position without clearance. After its heavy cast-iron mountings broke it was determined that the probable cause was stress caused by thermal expansion of the shaft. Determine the maximum force that could result from the estimated temperature change of 30°F. The material was steel with $C_t = 6.5 \times 10^{-6}$ in./in./°F and $E = 30 \times 10^6$ psi.

metric

The 100 mm shaft of a gyrating flour sifting machine was held in position without clearance. After its heavy cast-iron mountings broke it was determined that the probable cause was stress caused by thermal expansion of the shaft. Determine the maximum force that could result from the estimated temperature change of 20°C. The material was steel with $C_t = 1.30 \times 10^{-5}$ m/m/°C and $E = 210$ GPa.

Solution

customary

$$P = C_t(\Delta T)AE$$
$$P = 6.5 \times 10^{-6}(30°)(\pi 2^2)(30 \times 10^6)$$
$$P = 73,500 \text{ lb}$$

metric

$$P = C_t(\Delta T)AE$$
$$P = 1.30 \times 10^{-5}(20°C)$$
$$\times (\pi 0.050^2 \, m)(210 \, GPa)$$
$$P = 429\,000 \text{ N}$$

lesson: In design you must provide room for expansion or contraction due to temperature changes.

2-7 ALLOWABLE STRESS AND FACTOR OF SAFETY

Allowable stress, sometimes called working stress, is the maximum stress to which it is considered safe to subject a member in service. In many cases the allowable stress permitted is specified by governmental or other agencies in the form of codes or directives controlling different types of applications. The values are based on experiment and experience. In other cases company engineering departments may determine the allowable stress to be used. Otherwise it is up to the individual designer.

The ratio of the stress at which failure of a material will occur to the allowable stress is called the *factor of safety*. The term *factor of safety* is somewhat misleading. For example, a factor of safety of 2.5 does not mean that the member will withstand $2\frac{1}{2}$ times the allowable stress. The factor of safety of 2.5 simply means that the largest load that can be safely placed on the member is one that develops a stress no greater than the stress that would cause failure divided by 2.5.

Failure of a material may occur in a number of different ways depending on the requirements of the design. In some cases failure is considered to occur only when the member breaks. In this case the factor of safety would be

$$FS = \frac{f_u}{F_a} \qquad (2-8)$$

where FS is the factor of safety, f_u is the tensile strength of the material, and F_a is the allowable stress.

In the majority of cases a member will have failed if it is stressed beyond

Table 2.1

Allowable Stresses in Tension for Structural Steels

Specified Minimum Yield Stress, ksi	Allowable Stress on Net Section Except at Pin Holes or Threaded Parts, ksi	
	$0.60\,f_y$*	$0.55\,f_y$*
36	22.0**	20.0++
42	25.2**	23.0+
46	27.6	25.0+
50	30.0**	27.0+
60	36.0**	33.0+
70	42.0	38.5
80	48.0	44.0
90	52.5**	48.0+
100	57.5**	53.0+

Source. Courtesy of United States Steel Corporation.
*The values listed for 90 and 100 ksi yield strength steels are based on USS "T-1" steel and are limited by the tensile strength; AISC limits the allowable to $0.50f_u$ and AASHO limits the allowable to $0.46f_u$.
**Value according to AISC specification.
+Value according to AASHO specification.
++Value according to AREA specification.

the yield point. In these cases the factor of safety is

$$FS = \frac{f_y}{F_a}$$
(2-9)

In many cases the factors of safety are not given directly but are given in terms of the percentage of the yield stress f_y or of the tensile strength f_u as in Table 2-1.

Example

customary

A chandelier weighing 600 lb is supported by a 1/4 in. diameter aluminum wire having a yield stress of 48,000 psi. Determine the actual factor of safety based on the yield stress.

metric

A chandelier weighing 300 kg is supported by a 7 mm diameter aluminum wire having a yield stress of 324 MPa. Determine the actual factor of safety based on yield stress.

Solution

customary

$$F_a = \frac{P}{A} = \frac{600 \text{ lb}}{(0.125)^2} = 12{,}200 \text{ psi}$$

$$FS = \frac{f_y}{F_a} = \frac{48{,}000 \text{ psi}}{12{,}000 \text{ psi}} = 3.93$$

metric

$$F_a = \frac{P}{A} = \frac{300 \text{ kg}(9.81) \text{ N}}{(0.000\,35 \text{ m})^2} = 76.5 \text{ MPa}$$

$$FS = \frac{f_y}{F_a} = \frac{324 \text{ MPa}}{76.5 \text{ MPa}} = 4.24$$

2-8 ACCOMPLISHMENT CHECKLIST

From your study of this chapter you should now

1. Understand the natures of stress and strain.

2. Understand the difference between tensile, compressive, and shearing stress.

3. Understand the relationship between stress and strain.

4. Understand the terms, *proportional limit, elastic limit, yield point, tensile strength*, and *percentage elongation*.

5. Know how to apply Hooke's law and the modulus of elasticity in solving problems to members of one or more cross sections.

6. Understand how temperature changes cause changes in dimensions of members and how to apply the coefficient of thermal expansion to determine those changes.

7. Understand how to determine the maximum force that can be exerted by a constrained body upon its supports when its temperature changes.

8. Understand the meaning of allowable stress.

9. Know the meaning of factor of safety and how it may have different bases in different applications.

2-9 SUMMARY

Stress is the resistance of the molecules of a member to being pulled apart by the application of an external load. It is measured by the relationship

$$f = \frac{P}{A} \tag{2-1}$$

where f is the stress, P the externally applied load, and A is the cross section of the member.

Stress is classified as *tensile, compressive,* or *shearing* depending on whether the externally applied force is respectively pulling on, pushing on, or acting perpendicularly to the axis of the member. *Strain* is the deformation of a part per unit of its length, breadth, or width. It is determined by dividing the total deformation by the total dimension of the part in the direction of the deformation.

$$\epsilon = \frac{\Delta}{L} \tag{2-2}$$

where ϵ is the strain, Δ is the total deformation, and L is the total length in the direction of the deformation. Stress and strain are related in accordance with Hooke's law which can be stated in equation form as

$$f = \epsilon E \tag{2-3}$$

where f is the unit stress, ϵ is the unit strain, and E is the modulus of elasticity. From this relationship is developed the equation for determining the total deflection of a member

$$\Delta = \frac{PL}{AE} \tag{2-4}$$

Strain in the direction of the axis of the forces' application is called *axial strain*, whereas strain perpendicular to the applied load is *lateral strain*. The ratio of lateral strain to axial strain is *Poisson's ratio*. Mathematically it is

$$u = \frac{\epsilon_\ell}{\epsilon_a} \tag{2-5}$$

When most materials undergo temperature changes they expand or contract. The amount of this change is given by the *coefficient of thermal expansion* as the amount of expansion per unit of length per degree change in temperature. To determine the change in dimension due to a change in temperature the following equation may be used

$$\Delta = C_t L (\Delta T) \tag{2-6}$$

where C_t is the coefficient of linear expansion, L the total length of the part, Δ the total change in length, and ΔT the degrees change in temperature. When a body is prevented from expanding or contracting during a temperature change it may exert great forces on the members that prevent movement. The maximum amount of this force would be

$$P = C_t(\Delta T)AE \tag{2-7}$$

Allowable stress is the maximum stress which is considered safe for a member in service. The ratio of the stress at which failure of a material will occur to the allowable stress is called the *factor of safety*. Failure may occur in different ways depending on the criteria which would render the member unusable for the purpose for which it was intended. In most cases the member will be considered to have failed if it undergoes permanent deformation (*permanent set*). In other instances it will not be considered to have failed until it breaks (ruptures). The stresses and the factor of safety are related by the following equations:

$$FS = \frac{f_u}{F_a} \tag{2-8}$$

and

$$FS = \frac{F_y}{F_a} \tag{2-9}$$

where FS is the factor of safety, f_u is the tensile strength, f_y is the yield strength, and F_a is the allowable stress.

PROBLEMS

Note: All problems in this book are given in both U.S. Customary and SI Metric systems of units. The prefix C will be used for problems in the Customary System and the prefix M for the Metric System. The dimensions used in the Metric problems are not just conversions of those of the equivalent Customary problem but are those which would be used in original design using the Metric system. For example, a 2-15/16 dia shaft is actually 74.6 mm in diameter. In the metric problem we would use the nearest standard size of metric shafting, which would be 75 mm.

STRESS AND STRAIN

stress problems

2-1C   A 10 gage (0.1350 dia) wire is used as a guy wire under 200 lb tension. Determine the stress in the wire.

2-1M   *A 3.5 mm steel wire used as a guy wire is under 800 N tension. Determine the stress in the wire.*

2-2C   A short 4 × 4 wood post used as a support under a summer cottage carries a load of 2500 lb. Determine the stress in the post.

2-2M   *A short 10 cm × 10 cm wood post used as a support under a summer cottage carries a 1 000 kg load. Determine the stress in the post.*

2-3C   You are designing a swing that may be used by both children and adults. You plan to use aluminum chain with a maximum allowable stress for this application of 1000 psi. The maximum load that can be anticipated for the swing is 800 lb. Determine the minimum cross-sectional area for any point in the chain.

2-3M   *You are designing a swing that may be used by both children and adults. You plan to use aluminum chain with a maximum allowable stress for this application of 7 000 kN/m². The maximum load that can be anticipated for the swing is 350 kg. Determine the minimum cross-sectional area for any point in the chain.*

2-4C   A short W12X53 steel column supports a load of 200,000 lb. Determine the stress in the column. See Appendix 2 for information on the column.

2-4M   *A 300 × 300 − 87 kg/m steel column supports a load of 90 000 kg. Determine the stress in the column. See Appendix 2 for information on the column.*

2-5C   Select the most economical wide flange shape from Appendix 2 to be used as a short column to support a load of 260,000 lb. The most economical member will be the lightest. The column is to be made of A36 steel with a maximum yield stress f_y of 32 ksi. Assume that the column will fail in compression.

2-5M *Select the most economical wide flange shape from Appendix 2 to be used as a short column to support a load of 120 000 kg. The most economical member will be the lightest. The column is to be made of A36 steel with a maximum yield stress f_y of 220 MPa. Assume that the column will fail in compression.*

2-6C The system in the figure is designed to support a maximum load L of 20,000 lb. Determine the stress in the 1-1/4 in. rod.

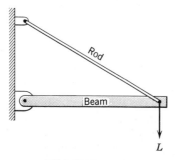

PROBLEM 2-6

2-6M *The system in the figure is designed to support a maximum load L of 20 000 kg. Determine the stress in the 30 mm rod.*

2-7C If the beam in Problem 2-6C is square and is made of pine wood having an allowable stress of 800 psi, what size should it be? Assume the beam to be short enough so that it would not bend.

2-7M *If the beam in Problem 2-6M is to be square and of pine wood having an allowable stress of 5.52 MPa, what size should it be? Assume that the beam is short enough that it would not bend.*

2-8C Determine the most economical (smallest weight) wide-flange beam from Appendix 2 for Problem 2-7C that will not be stressed over 8000 psi.

2-8M *Determine the most economical (smallest weight) wide-flange beam from Appendix 2 for Problem 2-7M that will not be stressed over 55 000 kN/m².*

2-9C   The 200 lb ball of the Foucault pendulum in the figure is supported by a high strength stainless steel wire having a diameter of 0.1055 in. (12 gage). Determine the stress in the wire when the ball is at the top of its swing (point A) and when it is at the bottom of its swing (point B). The radius *r* is 50 ft.

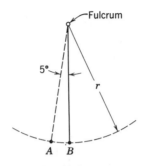

PROBLEM 2-9

2-9M   *The 100 kg ball of the Foucault pendulum in the figure is supported by a high strength stainless steel wire having a diameter of 3 mm. Determine the stress in the wire when the ball is at the top of its swing (point A) and when it is at the bottom of its swing (point B). The radius r is 15 cm.*

2-10C   If the maximum allowable compressive stress for a brick masonry wall is 200 psi and the masonry weighs 100 lb per cu ft, what is the highest wall that could be built of this brick without exceeding the allowable stress?

2-10M   *If the maximum allowable compressive stress for a brick masonry wall is 1380 kN/m² and the masonry weighs 1600 kg/m³, what is the highest wall that could be built of this brick without exceeding the allowable stress?*

strain problems

2-11C   The strain due to the loading of a shaft is 0.000007 in./in. Determine the deformation of the 3 ft long shaft.

2-11M   *The strain due to the loading of a shaft is 0.000 002 mm/mm. Determine the deformation of the shaft which is 1 m long.*

2-12C   The span of a bridge is 600 ft long. Due to an increase in temperature it undergoes a strain of 0.000315 in./in. How long is the span after the increase in temperature?

2-12M   *The span of a bridge is 200 m long. Due to a 30°C change in temperature it undergoes a strain of 0.000 189 m/m. How long is the span after the increase in temperature?*

2-13C   Under a load which develops a stress of 1750 psi in a polypropylene filament the strain is 0.01 in./in. Determine the total deformation in a filament 30 ft long.

2-13M   *Under a load which develops a stress of 12 000 kN/m² in a polypropylene filament the strain is 0.01 m/m. Determine the total deformation in a filament 15 m long.*

2-14C   A machine member under load has a total deformation of 0.027 in. The strain is 0.0006 in./in. Determine the length of the member.

2-14M   *A machine member under load has a total deformation of 0.068 cm. The strain is 0.000 272 m/m. Determine the length of the member.*

2-15C   A shaft mounted between thrust bearings will fail if the total deformation exceeds 0.0003 in. in the 4 ft distance between the bearings. Determine the maximum permissible strain.

2-15M   *A shaft mounted between thrust bearings will fail if the total deformation exceeds 0.000 07 mm in the 1.2 m distance between the bearings. Determine the maximum permissible strain.*

relationship of stress and strain (Hooke's law) problems

2-16C   The strain due to the loading of a shaft is 0.000007 in./in. The modulus of elasticity E is 30,000,000 psi. Determine the stress.

2-16M   *The strain due to the load on a steel shaft is 0.000 002 mm/mm. The modulus of elasticity is 210 GPa. Determine the stress.*

2-17C   The modulus of elasticity of an aluminum wire is 10×10^6 psi. The stress in it is 12,000 psi. Determine the strain.

2-17M   *The modulus of elasticity of an aluminum wire is 71×10^6 kN/m². The stress is 82 700 kN/m². Determine the strain.*

2-18C   In an experiment to determine the modulus of elasticity of a material it was found that a 1 in. × 1 in. test piece 10 in. long was deformed 0.004 in. by a 12,000 lb load. What was the modulus of elasticity of the material?

2-18M   *In an experiment to determine the modulus of elasticity of a material it was found that a 25 mm × 25 mm test piece 25 cm long was deformed 0.01 cm by a 53 kN load. What was the modulus elasticity of the material?*

2-19C   A 1/4-in. diameter nylon rod 20 ft long suspends a load of 490 lb. Its modulus of elasticity is 400,000 psi at 77°F. Determine the total elongation of the rod.

2-19M   *A 6 mm diameter nylon rod 6 m long suspends a load of 200 kg. Its modulus of elasticity is 2.7×10^6 kN/m² at 25°C. Determine the elongation of the rod.*

PROBLEM 2-20   (Courtesy Insley manufacturing.)

2-20C The bucket-tilting shaft of the backhoe in the illustration is 8 ft long and has a 3 in. diameter. Determine the strain and total deformation of the shaft at the allowable stress of 20,000 psi if $E = 30,000,000$ psi.

2-20M *The bucket-tilting shaft of the backhoe in the illustration is 2.5 m long and has a 75 mm diameter. Determine the strain and total deformation at the allowable stress of 138 MPa if $E = 210$ GPa.*

2-21C In a church he is designing an architect desires to have a 4-ton bronze cross suspended above the altar by an aluminum rod. Determine the minimum diameter that the rod must have it the allowable stress is 10,000 psi and the modulus of elasticity is 10,000,000 psi.

2-21M *In a church he is designing an architect desires to have a 4 000 kg bronze cross suspended above the altar by an aluminum rod. Determine the minimum diameter that the rod must have if the allowable stress is 69 000 kN/m² and the modulus of elasticity is 69×10^6 kN/m².*

2-22C A power plant boiler together with the tubing and fittings it supports weighs 940,000 lb. It is supported at each end by a support structure as shown in the figure. (a) Determine the size of the four wide-flange columns (vertical members) from Appendix 2 that would be required to support this load if the maximum allowable stress is 20,000 psi, assuming the columns would not buckle. (b) Determine the strain and deformation if A is 20 ft long and if $E = 29,000,000$ psi.

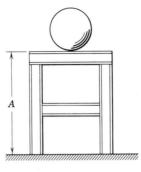

PROBLEM 2-22

2-22M   *The power plant boiler together with the tubing and fittings it supports weighs 400 000 kg. It is supported at each end by a support structure as shown in the figure. Determine the size of the four wide-flange columns (vertical members) that would be required to support this load if the maximum allowable stress is 140 000 kN/m², assuming they would not buckle. (b) Determine the strain and deformation if A is 6 m and if E = 200 × 10⁹ N/m².*

2-23C   Determine the total deformation of the vertical support shaft in the figure if the material is steel with a modulus of elasticity of 30,000,000 psi. Determine the stress in each of the three segments.

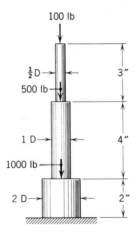

PROBLEM 2-23C

2-23M   *Determine the total deformation of the vertical support shaft in the figure if the material is steel with a modulus of elasticity of 210×10^6 kN/m². Determine the stress in each of the three segments.*

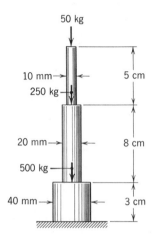

2-24C   Determine the total deformation of the vertical support shaft of Problem 2-23C if the 1/2-in. portion were nylon with E of 300,000 psi, the 1 in. portion were bronze with E of 15,000,000 psi, and the 2 in. portion were aluminum with E of 10,000,000 psi.

2-24M   *Determine the total deformation of the vertical support shaft of Problem 2-23M if the 10 mm portion were nylon with an E of 2.1×10^6 kN/m², the 20 mm portion were bronze with an E of 105×10^6 kN/m², and the 40 mm portion were aluminum with an E of 69×10^6 kN/m².*

2-25C   In many instances the weight of the supporting member itself is a factor in the design. For example, suppose a 20 lb instrument package were to be lowered into the ocean by an 11 gage (0.1205) stainless steel wire weighing 0.284 lb/in.³ and having a yield strength of 50,000 psi. Neglecting the buoyancy effects, determine how far the instrument package could be lowered without exceeding the yield strength assuming elastic deformation up to the yield strength. How much wire would be required for the job?

45

2-25M   *In many instances the weight of the supporting member itself is a factor in the design. For example, suppose a 10 kg instrument package were to be lowered into the ocean by a 3 mm (diameter) stainless steel wire weighing 7860 kg/m³ and having a yield strength of 350 000 kN/m². Neglecting the buoyancy effects, determine how far the instrument package could be lowered without exceeding the yield strength assuming elastic deformation up to the yield strength. How much wire would be required for the job?*

2-26C   The figure in view (a) shows a system of two rods with a zero load. View (b) shows the same system with $W = 10,000$ lb. The rods are 1/4-in. diameter steel with an E of 30,000,000 psi. $\overline{A'B'} = \overline{B'C'} = 10$ ft. Calculate the stress, strain, and elongation of the rods for the conditions in view (b). Also determine the angle θ and d.

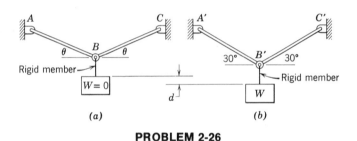

(a)   (b)

PROBLEM 2-26

2-26M   *The figure in view (a) shows a system of two rods with a zero load. View (b) shows the same system with $W = 4000$ kg. The rods are 6 mm in diameter, steel, and with $E = 200 \times 10^6$ kN/m². $\overline{A'B'} = \overline{B'C'} = 3$ m. Calculate the stress, strain, and elongation of the rods for the conditions of view (b). Also determine the angle θ and d.*

2-27C   Determine the minimum sizes of members for the crane system in the figure. The round support rod is wrought iron with an f_y of 45,000 psi. The square wood beam is made of fir with an allowable stress of 750 psi. The load W is 8000 lb and is suspended by a high-strength steel wire having an f_y of 180,000 psi. Design on the basis of yield strength for the rod and wire and allowable stress for the wood beam.

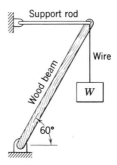

PROBLEM 2-27

2-27M   *Determine the minimum sizes of members for the crane system in the figure. The round support rod is wrought iron with an f_y of 310 MPa. The square wood beam is made of fir with an allowable stress of 5.20 MPa. The load W is 3500 kg and is suspended by a high-strength steel wire having an f_y of 1240 MPa. Design on the basis of yield strength for the rod and wire and allowable stress for the wood beam.*

Hooke's law problems

2-28C   A piece of steel strained axially 0.00067 in./in. is deformed laterally 0.00020 in./in. Determine Poisson's ratio for the steel.

2-28M   *A piece of steel strained axially 0.000 67 mm/mm is deformed laterally 0.000 20 mm/mm. Determine Poisson's ratio for the steel.*

2-29C   A 3 in. diameter shaft when subjected to an axial load increases in diameter to 3.0003 in. If the shaft was originally 10 in. long, what is its final length? The Poisson's ratio of the material is 0.3.

2-29M   *A 75 mm diameter shaft when subjected to an axial load increases in diameter to 76.2 mm. If the shaft was originally 250 mm long, what is its final length? The Poisson's ratio of the material is 0.3.*

2-30C   The pressure on objects submerged in sea water increases at the rate of 0.442 psi per foot of depth below the surface. Determine the dimensions of a cube of aluminum (Poisson's ratio = 0.33) at a depth of 20,000 ft if it was exactly 2 in. on a side when it was at the surface. $E = 10.1 \times 10^6$ psi.

2-30M   *The pressure on objects submerged in sea water increases at the rate of 10 000 Pa per meter of depth below the surface. Determine the dimensions of a cube of aluminum (Poisson's ratio = 0.33) at a depth of 6000 m if it was exactly 5 cm on a side when at the surface. $E = 69 \times 10^9$ Pa.*

temperature effects problems

2-31C   Determine the elongation of a steel rod 30 ft long due to a change in temperature of 80°F. Use the value of the coefficient of thermal expansions C_t in Appendix 4.

2-31M   *Determine the elongation of a steel rod 10 m long due to a change in temperature of 50°C. Use the value of the coefficient of thermal expansion C_t in Appendix 4.*

2-32C   A piece of wire 50 ft long changes length by 0.270 in. due to a temperature change of 60°F. Determine the coefficient of linear expansion. What is the material?

2-32M   *A piece of wire 20 m long changes length by 2.92 cm due to a temperature change of 90°C. Determine the coefficient of linear expansion. What is the material?*

2-33C   It is desired to fit a 3.000 in. shaft into a 2.999 in. hole. To do this the shaft is to be cooled until its diameter is 2.996 in. at which diameter it can be easily slipped into the hole. Determine how much cooler the shaft must be than the member containing the hole. Both parts are steel.

2-33M   *It is desired to fit a 75.000 mm shaft into a 74.975 mm hole. To do this the shaft is to be cooled until its diameter is 74.899 mm at which diameter it can be easily slipped into the hole. Both parts are steel.*

2-34C   In the system shown in the figure, A is 3 in., B is 10 in., C is 15 in., and D is 0.006 in. Determine the increase in temperature required to close the gap D. The coefficient of expansion of the left-hand post is zero.

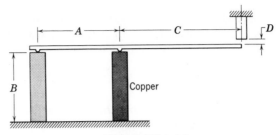

PROBLEM 2-34

2-34M *In the system shown in the figure, A is 75 mm, B is 250 mm, C is 375 mm, and D is 0.15 mm. Determine the increase in temperature required to close the gap D. The coefficient of expansion of the left-hand post is zero.*

2-35C The same as Problem 2-34C except that the right-hand post is steel.

2-35M *The same as Problem 2-34M except that the right-hand post is steel.*

2-36C A W6X7.85 aluminum H-beam 2 ft long is rigidly held between two immovable supports. Determine the force it exerts on the supports when its temperature is increased by 70°F. E for the aluminum is 10.1×10^6 psi.

2-36M *A $150 \times 150 - 10.6$ aluminum H-beam 1 m long is rigidly held between two immovable supports. Determine the force it exerts on the supports when the temperature is increased by 125°C. E for the aluminum is 69 GPa.*

2-37C It is desired to mount the machinery steel support bar in the figure so that it will always be in tension at ordinary temperatures. At room temperature 70°F

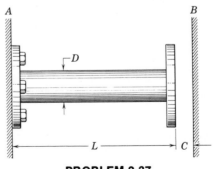

PROBLEM 2-37

the diameter D is 3 in., L is 24 in., and C is 0.006 in. The bar is attached to wall A and is to be heated until it touches wall B, to which it will then be attached. At what temperature will the rod touch wall B? When the temperature returns to 70°F what force will the rod be exerting on its supports?

2-37M *It is desired to mount the machinery steel support bar in the figure so that it will always be in tension at ordinary temperatures. At room temperature 18°C the diameter D is 75 mm, L is 0.6 m, and C is 0.15 mm. The bar is attached to wall A and is to be heated until it touches wall B, to which it will then be attached. At what temperature will the rod touch wall B? When the temperature returns to 18°C what force will the rod be exerting on its supports?*

2-38C A round 3003-H18 aluminum tube exactly 10 ft long at 32°F has an outside diameter of 10 in. and a wall thickness of 0.500 in. What is its height when supporting a 70,000 lb load at 120°F?

2-38M *A round 3003-H18 aluminum tube exactly 3 m long at 0°C has an outside diameter of 25 cm and a wall thickness of 1.0 cm. What is its height when supporting a 30 000 kg load at 50°C?*

factor of safety problems

2-39C Determine the factor of safety of a structural steel member that has a minimum yield stress of 50,000 psi and is designed using an allowable stress of 30,000 psi.

2-39M *Determine the factor of safety of a structural steel member that has a minimum yield stress of 350 000 kN/m^2 and is designed using an allowable stress of 210 000 kN/m^2.*

2-40C If the minimum yield stress of a material is 46,000 psi, what is the allowable stress for a factor of safety of 1.8?

2-40M *If the minimum yield stress of a material is 320 MPa, what is the allowable stress for a factor of safety of 1.8?*

2-41C   A W24X100 column carries a load of 700 kips. The A36 steel from which it is made has a specified minimum yield strength of 36 kips. Determine the factor of safety based on yield strength.

2-41M   *A 600 × 300 − 137 kg/m column carries a load of 320 000 kg. The A36 steel from which it is made has a specified minimum yield strength of 250 MPa. Determine the factor of safety based on yield strength.*

2-42C   In the figure, all members are free to pivot at the joints *A*, *B*, *C*, and *D*. Member *AB* is 0.500 diameter 2014-T6 aluminum, member *BC* is 0.500 diameter A441 steel, member *AC* is 6 in. × 6 in. Douglas fir, member *AD* is 15/32 diameter A572-Grade 65 steel with an f_y of 65,000 psi, and member *CD* is 0.300 copper round with an f_y of 145,000 psi. Determine the factor of safety of the system based on the yield strengths of the metals and the maximum compressive strength of the wood and determine which member is the controlling member.

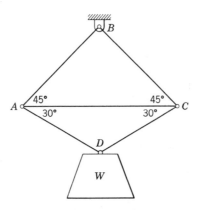

PROBLEM 2-42

2-42M   *In the figure, all members are free to pivot at the joints A, B, C, and D. Member AB is 15 mm diameter 2014-T6 aluminum, member BC is 13 mm diameter A441 steel, member AC is 150 mm × 150 mm Douglas fir, member AD is 12 mm diameter A572-Grade 65 steel with an f_y of 448 MPa, and member CD is 9 mm copper round with an f_y of 1 GPa. Determine the factor of safety of the system based on the yield strengths of the metals and the maximum compressive strength of the wood and determine which member is the controlling member.*

CHAPTER
OBJECTIVES ————————————————

The purpose of your study of this chapter is to

1. Come to an understanding of shearing forces and stresses.

2. Learn the relationship between tension, compression, and shear as they occur together in bodies.

3. Learn how to determine the forces and stresses in statically indeterminate bodies under axial loads.

4. Understand the strain-energy concept and how to apply it to determine stresses in members of machines and structures.

5. Learn how the shapes of members can cause the concentration of stresses which can lead to failures.

CHAPTER ─────────────────────── 3
TENSION, COMPRESSION,
AND SHEAR

In our study of stress and strain we came to a general understanding of tension and compression. In this chapter we will use the terms *direct tension* and *direct compression* and *direct shear*. Direct tension occurs in a two-force member when the forces are pulling on it in the axial direction. Similarly, direct compression occurs in a two-force member when the forces are pushing on it in the axial direction. Direct shear is described in section 3-1 below.

3-1 DIRECT SHEAR

When a force acts on a member in such a manner that it tends to cause the member to separate along a plane parallel to the applied force, it is called a *shearing force*. The internal resisting force is called *shearing stress* or more commonly, just *shear*. Its quantity is given in psi or Pa. In Figure 3-1(a) we see a rigidly held member with a cutting blade just touching it as it moves downward with a force P. Figure 3-1(b) shows the member being separated as the cutting blade moves downward. This is an example of *direct shear* in which all of the forces acting in the system are acting parallel to the direction of motion of the cutting blade. The shear in this case would be the force P divided by the cross-sectional area being separated (sheared). Mathematically, it is

$$f_v = \frac{P}{A_s} \tag{3-1}$$

where f_v is the shear in psi or Pa, P is the load or applied force, and A_s is the area sheared.

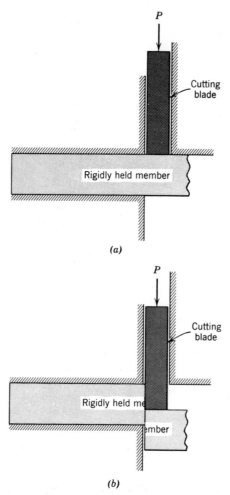

(a)

(b)

FIGURE 3-1

Example

customary

If, in Figure 3-1, the rigidly held member is 1/2-in. thick and 12 in. long, what force would be required to shear the end off if its ultimate strength in shear is 58,000 psi?

metric

If, in Figure 3-1, the rigidly held member is 1 cm thick and 30 cm long, what force would be required to shear the end off if its ultimate strength in shear is 400 MPa?

Solution

customary

Solving Equation 3-1 for P gives us

$$P = f_{us}A_s$$

where P = the required force,
f_{us} = the ultimate shearing strength
= 58,000 psi
A_s = the sheared area
= 1/2 in. (12 in.) = 6 in.^{2}
Substituting these values in the equation yields

$$P = 58,000 \text{ psi } (6 \text{ in.}^2) = 348,000 \text{ lb}$$
$$= 174 \text{ tons}$$

metric

Solving Equation (3-1) for P gives us

$$P = f_{us}A_s$$

where P = the required force,
f_{us} = the ultimate strength in shear
= 400 000 kPa,
A_s = the sheared area
= 0.01 m (0.30) = 0.003 m^2.
Substituting these values in the equation yields

$$P = 400 \text{ MPa } (0.003 \text{ m}^2) = 1\ 200 \text{ kN}$$

The foregoing examples involved *single shear*, shear in which only one area was in shear. Figure 3-2 shows an example of *double shear* where there are shear areas at both A and B. The shear area in each case is the cross-sectional area of the rivet as shown at the right.

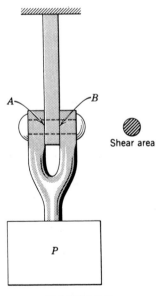

Shear area

FIGURE 3-2

Example

customary

Determine the shearing stress in the rivet of the clevis joint shown in Figure 3-2 if the rivet is 0.5 in. in diameter. $P = 2000$ lb.

metric

Determine the shearing stress in the rivet of the clevis joint shown in Figure 3-2 if the rivet is 10 mm in diameter. $P = 1\,000$ kg.

Solution

customary

Since the rivet is in double shear, the area under the shearing load is twice the cross-sectional area of the rivet or

$$A_s = 2\frac{\pi D^2}{4} = 2\frac{\pi(0.5 \text{ in.})^2}{4} = 0.392 \text{ in.}^2$$

Substituting this value in Equation 3-1 we get

$$f_v = \frac{P}{A_s} = \frac{2000 \text{ lb}}{0.392 \text{ in.}^2} = 5100 \text{ psi}$$

metric

Since the rivet is in double shear, the area under the shearing load is twice the cross-sectional area of the rivet or

$$A = 2\frac{\pi D^2}{4} = 2\frac{\pi(0.01 \text{ m})^2}{4}$$
$$= 1.57 \times 10^{-4} \text{ m}^2$$

Substituting this value in Equation 3-1 we get

$$f_v = \frac{P}{A_s} = \frac{1\,000 \text{ kg } (9.81 \text{ N/kg})}{1.57 \times 10^{-4} \text{ m}^2} = 62.5 \text{ MPa}$$

Just as there is proportionality between stress and strain in compression and tension, there is a like proportionality in shear. The constant of proportionality for shear is the *shear modulus of elasticity*. It is applied in the same manner as the modulus of elasticity for tension and compression.

$$f_v = G\epsilon \tag{3-2}$$

where f_v is the shearing stress, G is the shear modulus of elasticity, and ϵ is the strain.

3-2 RELATIONSHIP BETWEEN TENSION, COMPRESSION, AND SHEAR

Up to this point we have been studying direct stresses, sometimes called pure stresses. These stresses are simple, one-directional stresses which act in direct opposition to the applied forces that cause them. In many cases of simple

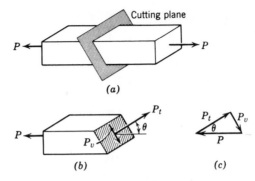

(a)

(b)   (c)

FIGURE 3-3

one-directional loading these are the only stresses that need be considered. In the following discussion it will be shown that there are cases of simple axial loading where it is necessary to consider shear stresses induced by axial loads. To determine these stresses it is necessary to develop relationships between compressive, tensile, and shear stresses.

To start, lets look at Figure 3-3. Here we have a rectangular bar which is held in tension by the forces P. Suppose that we were to pass a cutting plane through the bar as shown in (a) and then draw a free-body diagram of one end as in (b). It is seen that it takes two direct forces to provide equilibrium with the force P. The one acting perpendicular to the cut surface we will call P_t because it represents a direct tensile force. Similarly, we will denote the other force which is parallel to the surface (shearing) P_v. From the force diagram (c), because the angle between P_t and P_v is a right angle,

$$P_t = P \cos \theta \quad \text{and} \quad P_v = P \sin \theta$$

Dividing both equations by A, the cross-sectional area of the bar perpendicular to the axis yields

$$\frac{P_t}{A} = \frac{P}{A} \cos \theta \quad \text{and} \quad \frac{P_v}{A} = \frac{P}{A} \sin \theta$$

Since $A = A_{cp} \cos \theta$, where A_{cp} is the area of the cut plane, we can write the equations as

$$\frac{P_t}{A_{cp} \cos \theta} = \frac{P}{A} \cos \theta \quad \text{and} \quad \frac{P_v}{A_{cp} \cos \theta} = \frac{P}{A} \sin \theta$$

or

$$\frac{P_t}{A_{cp}} = \frac{P}{A} \cos^2 \theta \quad \text{and} \quad \frac{P_v}{A_{cp}} = \frac{P}{A} \sin \theta \cos \theta$$

57

and since force divided by the area equals the stress, we have

$$f_t = f_a \cos^2 \theta \quad \text{and} \quad f_v = f_a \sin \theta \cos \theta \qquad \text{(3-3) and (3-4)}$$

where f_t is tensile stress normal to (perpendicular to) the cut area, f_v is the shearing stress parallel to the cut surface, and f_a is the axial stress in the member (over the cross section normal to the axis of the member). The relationship for compressive stress is the same as that for tensile or

$$f_c = f_a \cos^2 \theta \qquad \text{(3-5)}$$

where f_c is the compressive stress over the cut area.

An important point concerning the preceding discussion is that, while a member is undergoing direct tensile or compressive loading, there are, in addition to the direct tensile and compressive stresses, stresses induced in different directions within the material. These may be analyzed by passing imaginary cutting planes through the member. The tensile or compressive stresses on these imaginary planes are of little significance, however, the shearing stress can be critical. Analysis of Equation 3-4 reveals that f_v will be at its maximum when θ is 45°. Thus it is found that when many machine or structural members fail under direct tension or compression, the plane of the failure will be at a 45° angle with their axes. An example is the standard tension test. The final fracture of steel in this test (as seen in Figure 2-5) has the shape of a cone, the surface of which makes an angle of 45° with the axis of the test piece. Similarly, cast iron under compression in the standard compression test will fail along a plane at 45° with the axis of the test piece.

When θ is 45°, $(\cos \theta \sin \theta) = 0.500$. Thus it is apparent that a material having an ultimate shear strength, f_s, of less than half its ultimate compressive or tensile strength f_a, can be expected to fail by shearing. For example, cast iron with an ultimate compressive strength of 75,000 psi (520 MPa) has an ultimate shearing strength of 30,000 psi (210 MPa) which is less than half of the compressive strength. It fails in shear.

Example

customary

A piece of 2 in. diameter shafting is pulled on by a 20,000 lb force. Determine the tensile and shearing stresses on a plane cutting the shafting at an angle θ of 45°.

metric

A piece of 50 mm shafting is pulled on by a 90 kN force. Determine the tensile and shearing stresses on a plane cutting the shafting at an angle θ of 45°.

Solution

customary

$$f_a = \frac{P}{A}, \qquad P = 20{,}000 \text{ lb},$$

$$A = \frac{\pi d^2}{4} = \frac{\pi 2^2}{4} = \pi \text{ in.}^2$$

$$f_a = \frac{20{,}000 \text{ lb}}{\pi \text{ in.}^2} = 6370 \text{ psi}$$

$$f_t = f_a \cos^2 \theta = 6370 \text{ psi (cos } 45°)^2$$
$$= 3180 \text{ psi}$$

$$f_v = f_a \sin \theta \cos \theta$$
$$= 6370 \text{ psi sin } 45° \cos 45° = 3180 \text{ psi}$$

metric

$$f_a = \frac{P}{A}, \qquad P = 90 \text{ kN},$$

$$A = \frac{\pi d^2}{4} = \frac{\pi (0.05 \text{ m})^2}{4} = 1.96 \times 10^{-3} \text{ m}^2$$

$$f_a = \frac{90 \text{ kN}}{1.96 \times 10^{-3} \text{ m}^2} = 45.9 \text{ MPa}$$

$$f_t = f_a \cos^2 \theta = 45.9 \text{ MPa (cos } 45°)^2$$
$$= 22.9 \text{ MPa}$$

$$f_v = f_a \sin \theta \cos \theta$$
$$= 45.9 \text{ MPa (sin } 45°)(\cos 45°)$$
$$= 22.9 \text{ MPa}$$

The value $\theta = 45°$ used in the foregoing examples gives us the maximum shearing stress for a member in direct tension. Thus, when the tension test sample fails with rupture at the 45° angle, the failure has come at maximum shear conditions. Further development of the relations between tension, compression, and shear, particularly dealing with combined stresses, will be discussed later in connection with the practical application of that knowledge.

3-3   STATICALLY INDETERMINATE MEMBERS UNDER AXIAL LOADS

In many systems composed of individual or composite members which are loaded axially it is not possible to determine the forces, stresses, and deformations in those members by the techniques of static equilibrium. Such members and systems are said to be *statically indeterminate.* The forces in these members cannot be determined because there are more unknowns in the problem than the number of independent equations that can be determined by the methods of statics. From our algebra we know that we must have at least as many independent equations relating the unknowns as we have unknowns. To solve problems involving statically indeterminate members it is necessary to use equations relating the elastic qualities of the material with those of statics.

Example

customary	metric
Figure 3-4 shows a bimetal powdered metal compression member. $L = 4$ in. Determine the deformation of the member due to the uniform load which is 40,000 lb. $D_1 = 2$ in. $D_2 = 3$ in.	*Figure 3-4 shows a bimetal powdered metal compression member. $L = 100$ mm, $D_1 = 50$ mm, and $D_2 = 75$ mm. Determine the deformation of the member due to the uniform load, which is 180 kN.*

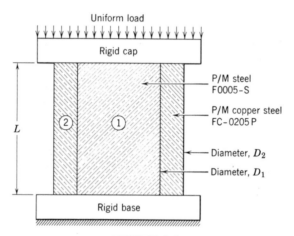

FIGURE 3-4

Solution

customary	metric
We have only one equation available to us by statics; that is from equilibrium of the rigid cap in the vertical direction. This gives us	*We have only one equation available to us by statics; that is from equilibrium of the rigid cap in the vertical direction. This gives us*
$$P_1 + P_2 = 40{,}000 \text{ lb}$$	$$P_1 + P_2 = 180 \text{ kN}$$
where P_1 is the portion of the load taken by the inside cylinder and P_2 is the portion of the load taken by the outside cylinder.	*where P_1 is the portion of the load taken by the inside cylinder and P_2 is the portion of the load taken by the outside cylinder.*

In order to solve the problem we must have another relationship equating P_1 and P_2. For this we look for a relationship in the elastic qualities of the materials. Study of the figure shows that parts 1 and 2 are equally compressed by the uniform load. Thus

$$\Delta_1 = \Delta_2$$

and since from Equation 2-4,

$$\Delta = \frac{PL}{AE}$$

$$\frac{P_1 L_1}{A_1 E_1} = \frac{P_2 L_2}{A_2 E_2}$$

Of the unknowns in the above equation we know that $L_1 = L_2 = 4$ in. We can determine that $A_1 = 3.14$ in.^{2} and $A_2 = 7.07$ in.^{2}

From Appendix 3 $E_1 = 19 \times 10^6$ psi and $E_2 = 13 \times 10^6$ psi. Substituting these values in the above equation gives us

$$\frac{P_1\,(4\text{ in.})}{3.14\text{ in.}^2(19 \times 10^6)} =$$

$$\frac{P_2\,(4\text{ in.})}{7.07\text{ in.}^2\,(13 \times 10^6)}$$

Solving the above reveals

$$P_1 = 0.649 P_2$$

We then substitute this value of P_1 in our first equation

$$0.649 P_2 + P_2 = 40{,}000\text{ lb}$$
$$P_2 = 24{,}300\text{ lb}$$

We can now determine Δ_2 and since $\Delta_1 = \Delta_2$ this will give us the deformation of the whole member.

In order to solve the problem we must have another relationship equating P_1 and P_2. For this we look for a relationship in the elastic qualities of the materials. Study of the figure shows us that parts 1 and 2 are equally compressed by the uniform load. Thus,

$$\Delta_1 = \Delta_2$$

and since from Equation 2-4,

$$\Delta = \frac{PL}{AE}$$

$$\frac{P_1 L_1}{A_1 E_1} = \frac{P_2 L_2}{A_2 E_2}$$

Of the unknowns in the above equation we know that $L_1 = L_2 = 0.1$ m. We can determine that $A_1 = 1.96 \times 10^{-3}$ m^2 and $A_2 = 4.42 \times 10^{-3}$ m^2.

From Appendix 3 $E_1 = 131 \times 10^9$ Pa and $E_2 = 89.6 \times 10^9$ Pa. Substituting these values in the above equation gives us

$$\frac{P_1\,(0.1\text{ m})}{1.96 \times 10^{-3}\text{ m}^2\,(131 \times 10^9\text{ Pa})} =$$

$$\frac{P_2\,(0.1\text{ m})}{4.42 \times 10^{-3}\text{ m}^2\,(89.6 \times 10^9\text{ Pa})}$$

Solving the above reveals

$$P_1 = 0.648 P_2$$

We then substitute this value of P_1 in our first equation

$$0.648 P_2 + P_2 = 180\text{ kN}$$
$$P_2 = 109\text{ kN}$$

We can now determine Δ_2, and because $\Delta_1 = \Delta_2$, this will give us the deformation of the whole member.

$$\Delta_2 = \frac{P_2 L_2}{A_2 E_2} = \frac{24{,}300 \text{ lb (4 in.)}}{7.07 \text{ in.}^2 (13 \times 10^6)}$$

$$= 0.00106 \text{ in.}$$

$$\Delta_2 = \frac{P_2 L_2}{A_2 E_2}$$

$$= \frac{109 \text{ kN (0.1 m)}}{4.42 \times 10^{-3} \text{ m}^2 (89.6 \times 10^9 \text{ Pa})}$$

$$= 2.75 \times 10^{-5} \text{ m}$$

$$= 0.027 \, 5 \text{ mm}$$

3-4 STRAIN ENERGY

To this point in our study we have been assuming that our loads were static; either they were dead loads or the loads were applied very slowly. In many instances loads are not static but are moving. It is important to be able to determine the effects of these moving (dynamic) loads. One way is to apply the concept of strain energy. Figure 3-5 illustrates the concept of strain energy.

In (a) we see a mass (moving or stationary) which is in contact with the column but is exerting no force on it. In (b) it has moved down a distance Δ and is exerting a force P on the column. During this movement the mass has given up its energy which now exists in the column in the form of elastic potential energy, which we call *strain energy*. To determine the relationship between the energy in the mass in (a) to the strain energy in the column in (b) we will look at the load-deformation diagram (c). Considering the column to be perfectly elastic, the deformation increases uniformly from 0 to Δ as the load imposed by the mass increases from 0 to P. The energy absorbed by the

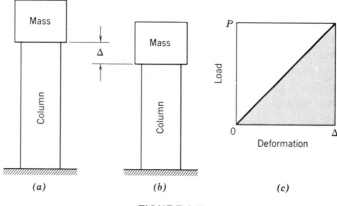

FIGURE 3-5

column during this deformation is represented by the shaded area under the curve.

$$U_{st} = \frac{1}{2} P\Delta$$

where U_{st} is the strain energy or since

$$\Delta = \frac{PL}{AE}$$

$$U_{st} = \frac{1}{2} \frac{P^2 L}{AE} \quad \text{or} \quad \frac{P^2}{A^2} \frac{LA}{2E}$$

and since

$$\frac{P}{A} = f \qquad U_{st} = \frac{f^2 LA}{2E} \tag{3-6}$$

The maximum strain energy that can be absorbed by a member is given by

$$U_{ust} = \frac{f_e^2 LA}{2E} \tag{3-7}$$

where f_e is the *stress at the elastic limit* of the material.

The ability of a material to absorb energy is *resiliency*. The amount of energy that a material will absorb per unit volume by being stressed up to its elastic limit and which can be recovered from it is its *modulus of resiliency*. Its value is given by

$$E_R = \frac{f_e^2}{2E} \tag{3-8}$$

The following example illustrates how the concept of strain energy can be applied to a practical problem.

Example

customary

A tie rod is required to absorb a shock load. Determine which of the two designs in Figure 3-6 would best serve the purpose. The upper rod has a diameter equal to the nominal thread diameter throughout its length. The lower rod has

metric

A tie rod is required to absorb a shock load. Determine which of the two designs in Figure 3-6 would best serve the purpose. The upper rod has a diameter equal to the nominal thread diameter throughout its length. The

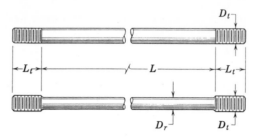

FIGURE 3-6

the portion between the threads turned down to the minor diameter of the threads. The minor diameter is the diameter at the bottom of the threads. The dimensions follow.

The nominal diameter (outside diameter) of the threads D_t is 1 in.
The minor diameter of the threads D_r is 0.8466 in.
The length of the threads L_t is 2 in.
The length of the rod L between the threads is 3 ft.
$E = 29,000,000$ psi and $f_e = 36,000$ psi

Note that for most steels the yield stress and elastic limit are the same.

lower rod has the portion between the threads turned down to the minor diameter of the threads. The minor diameter is the diameter at the bottom of the threads. The given dimensions follow.

The nominal diameter of the threads D_t is 25 mm.
The minor diameter of the threads D_r is 24.025 mm.
The length of the threads L_t is 50 mm.
The length of the rod L between the threads is 1 m.
$E = 200$ GPa and $f_e = 250$ MPa

Note that for most steels the yield stress and elastic limit are the same.

Solution

customary

For the upper rod the maximum stress in the threaded sections governs the load. The area of the minor diameter A_r is $\pi D_t^2/4$ or

$$A_r = \frac{\pi (0.8446)^2}{4} = 0.5603 \text{ in.}^2$$

$$P = A_r f_e = 0.5603 \text{ in.}^2 (36,000 \text{ psi})$$
$$= 20,200 \text{ lb}$$

metric

For the upper rod the maximum stress in the threaded sections governs the load. The area of the minor diameter A_r is $\pi D_r^2/4$ or

$$A_r = \frac{\pi (0.024\,025 \text{ m})^2}{4} = 0.000\,453 \text{ m}^2$$

$$P = A_r f_e = 0.000\,453 \text{ m}^2 (250 \text{ MPa})$$
$$= 68\,000 \text{ N}$$

The stress in the unthreaded section is

$$f = \frac{P}{A} = \frac{20{,}200 \text{ lb}}{\pi(1^2/4)} = 25{,}700 \text{ psi}$$

We can now find U_{st} for the upper rod to be

$$U_{st} = \frac{f_e^2 L_t A_t}{2E} + \frac{f^2 L A_r}{2E}$$

$$U_{st} = \frac{(36{,}000 \text{ psi})^2(4 \text{ in.})(0.5603 \text{ in.}^2)}{2\,(29{,}000{,}000 \text{ psi})}$$

$$+ \frac{(25{,}700 \text{ psi})^2(36 \text{ in.})(0.7854 \text{ in.}^2)}{2\,(29{,}000{,}000 \text{ psi})}$$

$$U_{st} = 372 \text{ in.-lb}$$

For the lower rod the minor diameter of the threaded portion of the rod and the diameter of the unthreaded portion of the rod are the same. Therefore, they both carry the same load at the same stress— the maximum stress f_e which is 36,000 psi. The strain energy for the lower rod will be

$$U_{ust} = \frac{(36{,}000 \text{ psi})^2(4 \text{ in.} + 36 \text{ in.})(0.5603 \text{ in.}^2)}{2\,(29{,}000{,}000 \text{ psi})}$$

$$= 501 \text{ in.-lb}$$

The lower rod is best for the purpose because it can absorb the most energy.

The area of the unthreaded rod is

$$A = \frac{\pi D_t^2}{4} = \frac{\pi\,(0.025 \text{ m})^2}{4} = 0.000\,491 \text{ m}^2$$

The stress in the unthreaded section is

$$f = \frac{P}{A} = \frac{68\,000 \text{ N}}{0.000\,491 \text{ m}^2} = 138 \text{ MPa}$$

We can now find U_{st} for the upper rod to be

$$U_{st} = \frac{f_e^2 L_t A_r}{2E} + \frac{f^2 L A_t}{2E}$$

$$U_{st} = \frac{(250 \text{ MPa})(0.050 \text{ m})(0.000\,453 \text{ m}^2)}{2\,(200 \text{ GPa})}$$

$$+ \frac{(138 \text{ MPa})^2(1 \text{ m})(0.000\,491 \text{ m}^2)}{2\,(200 \text{ GPa})}$$

$$U_{st} = 26.9 \text{ Pa}$$

For the lower rod the minor diameter of the threaded portion of the rod and the diameter of the unthreaded portion of the rod are the same. Therefore, they both carry the same load at the same stress—the maximum stress f_e which is 250 MPa. The strain energy for the lower rod will be

$$U_{ust} = \frac{(250 \text{ MPa})^2(1.1 \text{ m})(0.000\,453 \text{ m}^2)}{2\,(200 \text{ GPa})}$$

$$= 77.8 \text{ Pa}$$

The lower rod is best for the purpose because it can absorb the most energy.

3-5 IMPACT LOADS

In the previous section on strain energy we considered the ability of a structural or machine member to store energy, and in so doing we developed a relationship pertaining to the energy exchange between two bodies in terms of

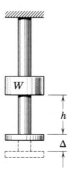

FIGURE 3-7

the elastic stress, volume, and modulus of elasticity of the member receiving the energy. It was stated by Equation 3-7, which can be rewritten as follows

$$U_{st} = \frac{f^2 L A}{2E} \qquad (3\text{-}7a)$$

where U_{st} and f are the developed strain energy and stress, respectively. Since U_{st} is the energy actually given up by one body to another body being strained and thus storing energy, we can apply it to a situation such as shown in Figure 3-7. Here we have a system where the weight W, when dropped from its position a distance h above the flange of the vertical rod, will cause the rod to deform the amount Δ. The strain energy contained in the vertical rod after deformation is equal to the energy given up by the weight due to its change in position, $W(h + \Delta)$. Substituting this in Equation 3-7 for U_{st} gives us

$$W(h + \Delta) = \frac{f^2 L A}{2E} \qquad (3\text{-}8)$$

We will solve the above equation for f because it is usually the stress developed in the body receiving the energy which is of interest.

Since $\Delta = fL/E$ we can substitute this in the above equation giving

$$W(h + \frac{fL}{E}) = \frac{f^2 L A}{2E} \qquad (3\text{-}8a)$$

solving this for f yields

$$f = \frac{W}{A} + \frac{W}{A}\sqrt{1 + \frac{2EAh}{WL}} \qquad (3\text{-}9)$$

Example

customary

Determine the maximum stress and deflection of a 3/4-in. diameter vertical shaft 8 in. long mounted on a rigid base caused by a 40 lb weight being dropped on it from a height of 6 in. The shaft is steel with an E of 29,000,000 psi.

metric

Determine the maximum stress and deflection of a 25 mm vertical shaft 200 mm long mounted on a rigid base caused by a 20 kg weight being dropped on it from a height of 150 mm. The shaft is steel with an E of 200 GPa.

Solution

customary

$W = 40$ lb, $A = \dfrac{\pi (3/4 \text{ in.})^2}{4} = 0.442 \text{ in.}^2,$

$h = 6$ in., $L = 8$ in.,

and

$$E = 29{,}000{,}000 \text{ psi}$$

Substituting these values in Equation 3-9 gives

$$f = \frac{40 \text{ lb}}{0.442 \text{ in.}^2} + \frac{40 \text{ lb}}{0.442 \text{ in.}^2}$$

$$\times \sqrt{1 + \frac{2\,(29{,}000{,}000 \text{ psi})(0.442 \text{ in.}^2)(6 \text{ in.})}{40 \text{ lb} (8 \text{ in.})}}$$

$$f = 62{,}800 \text{ psi}$$

and

$$\Delta = \frac{fL}{E} = \frac{62{,}800 \text{ psi} (8 \text{ in.})}{29{,}000{,}000 \text{ psi}} = 0.0173 \text{ in.}$$

metric

W must be in force units, so

$$W = 20 \text{ kg} (9.81) = 196 \text{ N},$$

$$A = \frac{\pi (0.025 \text{ m})^2}{4} = 0.000\,491 \text{ m}^2,$$

$$h = 0.15 \text{ m}, \qquad L = 0.20 \text{ m},$$

and

$$E = 200 \text{ GPa}.$$

Substituting these values in Equation 3-9 gives us

$$f = \frac{196 \text{ N}}{0.000\,491 \text{ m}^2} + \frac{196 \text{ N}}{0.000\,491 \text{ m}^2}$$

$$\times \sqrt{1 + \frac{(200 \text{ GPa})(0.000\,491 \text{ m}^2)(0.15 \text{ m})}{196 \text{ N} (0.20 \text{ m})}}$$

$$f = 346 \text{ MPa}$$

and

$$\Delta = \frac{fL}{E} = \frac{346 \text{ MPa} (0.20 \text{ m})}{200 \text{ GPa}}$$
$$= 0.000\,346 \text{ m} = 0.346 \text{ mm}$$

3-6 STRESS CONCENTRATION

We have been dealing with members as though they were of uniform dimensions and of flawless nature throughout. In most instances this is not the case. Wherever there are holes, notches, abrupt changes of contour, or even scratches or irregularities on the surface the stresses increase in the vicinity of these discontinuities. The most common example is illustrated in Figure 3-8 which shows a plate with a hole in it. The plate is in tension with the arrows indicating the distribution of stress, the length of the arrows being proportional to the intensity of stress. The stress at the edge of the hole is about three times that at the edge of the plate where the stress is uniformly distributed. At first this might seem to be cause for alarm when designing members with holes in them, but such is not the case. In the first place practically all such members are made of materials such as steel which upon reaching the yield stress will flow plastically. In our example the fibers nearest the hole will reach the point where they will flow plastically before those further from the hole. The plastic flow of the inner fibers relieves the excess stress which therefore need not be considered in design. The only instances where the stress concentration does become a problem are the cases where the material is too brittle to flow plastically under the stress, or when the load is applied so abruptly that plastic flow does not have sufficient time to occur. An example of the latter is the rapidly repeated loading of the reciprocating part of air hammers or nut drivers. Another case is that of the rapidly rotating shaft carrying a transverse load. It undergoes a stress reversal at the same frequency as the rotation of the shaft and therefore requires attention to stress concentrations.

The effect of stress concentrations is generally taken care of by applying multiplying factors to the standard stress equations. However, the proper use of such factors is quite involved and is beyond the scope of this textbook. Those desiring to further investigate stress concentration will find *Stress Concentration Factors* by R. E. Peterson (New York: Wiley, 1974) of interest.

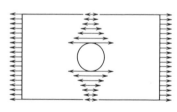

FIGURE 3-8

3-7 ACCOMPLISHMENT CHECKLIST

As a result of your studies do you now

1. **Understand the nature of shearing stress, the meaning of the term *direct shear*, and how to determine the direct shearing stress in and deformation of a member under the effect of a shearing force.**

2. **Understand the relationship between tension, compression, and shear in members and how to determine the tensile, compressive, and shear stresses on planes cut at angles to the axis of forces of a two-force member.**

3. **How to determine the forces, stresses, and deformations of statically indeterminate members under axial loading.**

4. **Understand the concept of strain energy and how to determine the amount of energy a member can absorb without failing.**

5. **Know the meaning of the term *resiliency* and understand the significance of the *modulus of resiliency*.**

6. **Know what *stress concentration* is and when it should be considered?**

3-8 SUMMARY

Direct tension occurs when a two-force member has forces pulling on it. *Direct compression* occurs when a two-force member has forces pushing on it.

A force tending to cause failure on a plane parallel with it is a *shearing force* and the resisting stress is the *shearing stress*. When the force and the shearing stress are in the same plane we have *direct shear*. Direct shearing stress can be determined by the relationship

$$f_v = \frac{P}{A_s} \tag{3-1}$$

where f_v is the shearing stress in psi or Pa, P is the load or applied force, and A_s is the area of the surface over which the shearing stress acts.

Where a single surface is under shear the condition is called *single shear*.

69

Where a member has two surfaces under shear such as the pin in a clevis joint the condition is one of *double shear.*

Shearing stress is related to strain by the *shear modulus of elasticity* by

$$f_v = G\varepsilon \qquad (3\text{-}2)$$

where G is the shear modulus of elasticity, f_v is the shear stress, and ε is the strain.

When a two-force member is placed under tension or compression stresses are induced in all directions. The relationships of these stresses over a surface cut by a plane at any angle to the axis of the member can be determined by the equations

$$f_u = f_a \cos^2 \theta \qquad \text{and} \qquad f_s = f_a \sin \theta \cos \theta \qquad (3\text{-}3) \text{ and } (3\text{-}4)$$

and

$$f_c = f_a \cos^2 \theta \qquad (3\text{-}5)$$

where f_u is the tensile stress perpendicular to the cut surface, f_s is shearing stress parallel to the cut surface, f_c is the compressive stress perpendicular to the cut surface, f_a is the axial stress over the area normal to the axis of the member, and θ is the angle a perpendicular to the cut surface makes with the axis of the member.

Members of which the forces and stresses cannot be determined by the methods of statics are said to be *statically indeterminate.* To solve problems involving statically indeterminate members it is necessary to use equations relating the elastic qualities of the material along with those of statics.

The *strain energy* concept is; the energy a moving load gives up to the member receiving the load shows up as the potential energy of deformation of the member, as long as the member is not stressed beyond the elastic limit. The amount of the strain energy is given by the equation

$$U_{st} = \frac{f^2 LA}{2E} \qquad (3\text{-}6)$$

where U_{st} is the strain energy, f is the stress in the energy absorbing member, L is its length, A is its cross-sectional area, and E is its modulus of elasticity.

The maximum energy the material can absorb within the elastic range is

$$U_{st} = \frac{f_e^2 LA}{2E} \qquad (3\text{-}7)$$

where f_e is the stress at the elastic limit of the material. The maximum amount of stress that a material can absorb per unit of volume is its *modulus of resiliency* E_R, the value of which is

$$E_R = \frac{f_e^2}{2E} \qquad (3\text{-}8)$$

Stress concentration is the increase in the level of stress in the vicinity of irregularities such as holes, notches, and sudden changes in contours. Stress concentration in materials that deform plastically is relieved by plastic deformation and consequently is not a problem for the designer. It must be considered where there are rapid repetitions or reversals of loading or where the material is brittle.

PROBLEMS

direct shear

3-1C   A 0.125 in. diameter rod embedded rigidly in a casting as shown in the figure carries a load P of 50 lb. Determine the shearing stress in the rod at the point where it enters the casting.

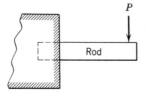

PROBLEM 3-1

3-1M   *A 3 mm diameter rod embedded rigidly in a casting as shown in the figure carries a load P of 20 kg. Determine the shearing stress in the rod at the point where it enters the casting.*

3-2C   What minimum diameter would the rod in Problem 3-1C have to be if it were made of 2024-T4 aluminum which has a minimum expected shear strength of 37,000 psi?

3-2M   *What minimum diameter would the rod in Problem 3-1M have to be if it were made of 2024-T4 aluminum which has a minimum expected shear strength of 255 MPa?*

3-3C   What maximum load could the 0.125 in. diameter rod in Problem 3-1C carry in shear if the rod were made of steel with an allowable shear stress of 25,000 psi?

3-3M   *What maximum load could the 3 mm diameter rod in Problem 3-1M carry in shear if the rod were made of steel with an allowable shear stress of 170 MPa?*

3-4C   It is desired to punch (blank) a 3 in. diameter hole through a 0.032 in. thick sheet of 2014-T6 aluminum which has an ultimate shear strength of 38,000 psi. Determine the force required.

3-4M   *It is desired to punch (blank) a 75 mm hole through a 0.800 mm thick sheet of 2014-T6 aluminum which has an ultimate shear strength of 262 MPa. Determine the force required.*

3-5C   What is the largest diameter of hole that can be punched through 0.25 in. steel plate by an 8-ton punch press if the ultimate shear stress of the steel is 45,000 psi?

3-5M   *What is the largest diameter of hole that can be punched through 6 mm thick steel plate by a punch press capable of delivering a force on the punch of 70 kN if the ultimate shear stress of the steel is 310 MPa?*

3-6C   It is desired to punch a pattern of four 0.500 in. diameter and one 1.500 diameter holes simultaneously in a 0.032 in. thick sheet of 3003-H12 aluminum which has an ultimate shear strength of 10,000 psi. Determine the force required.

3-6M   *It is desired to punch a pattern of four 10 mm diameter and one 40 mm diameter holes simultaneously in a 0.800 mm thick sheet of 3003-H12 aluminum which has an ultimate strength of 70 MPa. Determine the force required.*

3-7C   The figure shows the cross section of a simple riveted joint. The rivet is 0.500 in. diameter and has a yield strength f_{yv} in shear of 21,000 psi. What is the largest load P that can be placed on the joint without exceeding $0.40 f_{yv}$? Note that the rivet is in double shear. Friction of the joint is to be ignored.

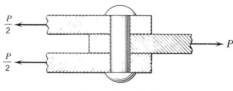

PROBLEM 3-7

3-7M   *The figure shows the cross section of a simple riveted joint. The rivet is 12 mm in diameter and has a yield strength in shear of 145 MN/m². What is the largest load P that can be placed on the joint without exceeding 0.40 f_{yv}? Note that the rivet is in double shear. Friction of the joint is to be ignored.*

3-8C   Determine the stress, strain, and deformation of the shaded area of the figure if t is 0.500 in., l is 3 in., P is 2000 lb, and G is 10,000,000 psi.

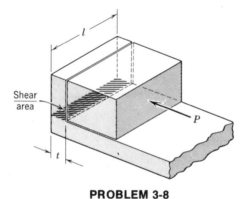

PROBLEM 3-8

3-8M   *Determine the stress, strain, and deformation of the shaded area of the figure if t is 15 mm, l is 75 mm, P is 20 000 N, and G is 70 GPa.*

3-9C   If P in Problem 3-8C were 10,000 lb, what would be the minimum thickness t for a yield stress in shear f_{ys} of 21,000 psi using a factor of safety N of 1.8?

3-9M   *If P in Problem 3-8M were 45 000 N, what would be the minimum thickness t for a yield stress in shear f_{ys} of 145 MPa using a factor of safety N of 1.8?*

TENSION, COMPRESSION, AND SHEAR

3-10C   Determine the shearing stress on the plane *mn* of the support if the force *P* is 40,000 lb, *t* is 1 in., the depth of the support, perpendicular to the paper, is 10 in., and the coefficient of friction C_{fs} is 0.3.

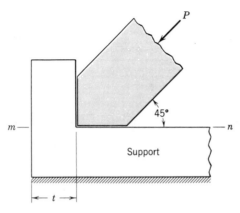

PROBLEM 3-10

3-10M   *Determine the shearing stress on the plane mn of the support if the force P is 180,000 N, t is 25 mn, the depth of the support, perpendicular to the paper, is 250 mm, and the coefficient of friction C_{fs} is 0.3.*

relationship between tension, compression, and shear

3-11C   The axial stress on an axially loaded steel bar of uniform cross section is 14,000 psi. Determine the tensile and shear stresses acting on a surface cut through the member at an angle θ of 30°.

3-11M   *The axial stress on an axially loaded steel bar of uniform cross section is 95 MPa. Determine the tensile and shear stresses acting on a surface cut through the member at an angle θ of 30°.*

3-12C   A rectangular beam is axially loaded. The axial stress produced by this loading is f_a. Determine the shear stresses on planes cut at angles of 44°, 45°, and 46° with the axis. At which angle is the shear stress a maximum?

3-12M *A rectangular beam is axially loaded. The axial stress produced by this loading is f_a. Determine the shear stresses on planes cut at angles of 44°, 45°, and 46° with the axis. At which angle is the shear stress a maximum?*

3-13C A square white oak post is to carry a maximum load of 20,000 lb. It is too short to buckle and will therefore fail in shear. Determine its minimum size if the allowable stress in shear is 490 psi.

3-13M *A square white oak post is to carry a maximum load of 10 000 kg. It is too short to buckle and will therefore fail in shear. Determine its minimum size if the allowable stress in shear is 3 MPa.*

3-14C The figure shows a 1 in. diameter rod attached to a block by an adhesive having a maximum shear strength of 400 psi. Determine the maximum load P that the system will take without failing.

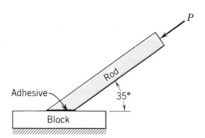

PROBLEM 3-14

3-14M *The figure shows a 25 mm diameter rod attached to a block by an adhesive having a maximum shearing strength of 2.75 MPa. Determine the maximum load P that the system will take without failing.*

3-15C The shear stress on a plane cut at an angle of 40° with the axis of an axially loaded 2 in. × 3 in. steel bar is 30,000 psi. Determine the stress normal to the shear plane and the axial load on the bar.

3-15M *The shear stress on a plane cut at an angle of 40° with the axis of an axially loaded 50 mm × 75 mm steel bar is 200 MPa. Determine the stress normal to the shear plane and the axial load on the bar.*

statically indeterminate members under axial load problems

3-16C   A piece of aluminum of 1.00 in. diameter 2 in. long is placed end to end with a piece of steel of 1.00 in. diameter and compressed between vice jaws exerting a force of 2500 lb. *E* for steel is 29,000,000 psi and *E* for aluminum is 10,000,000 psi. Determine the required original length of the steel in order that the total deformation of the steel be the same as that of the aluminum.

3-16M   *A piece of aluminum 25 mm in diameter and 5 cm long is placed end to end with a piece of steel of 25 mm diameter and is compressed between vice jaws exerting a force of 10 000 N. E for steel is 200 GPa and E for aluminum is 69 GPa. Determine the required original length of the steel in order that the total deformation of the steel be the same as that of the aluminum.*

3-17C   A wood post 8 in. square has a 1 in. diameter steel rod running through its center as shown in the figure. Both the wood post and steel rod were 4 ft long before being loaded. Determine their length when subjected to a 20,000 lb load and the stresses in each.

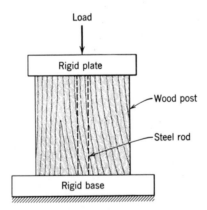

PROBLEM 3-17

3-17M   *A wood post 20 cm square has a 25 mm diameter steel rod running through its center as shown in the figure. Both the wood post and the steel rod were 1.2 m long before being loaded. Determine their length when subjected to a 10 000 kg load and the stresses in each.*

3-18C   The post shown in the figure has a load of 5 tons imposed on it by the crosshatched ring; both ends of the post are attached to a rigid frame. Without the 5 ton load there is no stress in the post. Determine the stresses in and the deformations of both the upper and lower parts of the post caused by the 5 ton load. The material of the post is steel with $E = 29,000,000$ psi. The height a is 3 ft, b is 5 ft, the diameter d_a is 2 in. and the diameter d_b is 4 in.

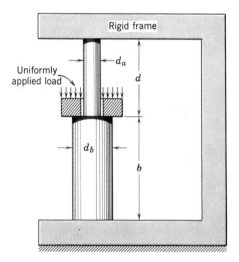

PROBLEM 3-18

3-18M   *The post in the figure has a load of 5 tons imposed on it by the crosshatched ring. Both ends of the post are firmly attached to a rigid frame. Without the 5 ton load there is no stress in the post. Determine the stresses in and the deformations of both the upper and lower parts of the post caused by the application of the 5 ton load. The material of the post is steel with $E = 200$ GPa. The height a is 1 m, b is 1.6 m, the diameter d_a is 50 mm and the diameter d_b is 100 mm.*

3-19C   The sign for Al's Place weighs 2000 lb with the weight evenly distributed over its length. It was to be supported by two 0.5 in. diameter rods 15 ft long (before loading) spaced 15 ft apart. Due to an error the left-hand rod came in copper, $(E = 17 \times 10^6$ psi) and the right-hand one came in aluminum $(E = 10 \times 10^6$ psi). Where between the aluminum and copper rods could a 0.5 in. diameter steel rod $(E = 29 \times 10^6$ psi) be placed so that the sign would hang horizontally?

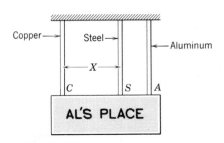

PROBLEM 3-19

3-19M *The sign for Al's Place weighs 1000 kg with the weight evenly distributed over its length. The supporting rods are both exactly 6 cm long and 10 mm diameter (before loading) and are spaced 6 m apart. Due to an error the left-hand rod came in copper (E = 117 GPa) and the right-hand rod came in aluminum (E = 69 GPa). Where between the copper and aluminum rods could a steel rod (E = 200 GPa) of 10 mm diameter be placed so that the sign would hang horizontally?*

3-20C In the figure you are looking at the edges of the three bars, A, B, and C. The bars have a uniform cross section of 1.5 in.². The steel pins D are 1.000 in. in diameter. The bars A and C are copper with an E of 18,000,000 psi and with the distance between the hole centers prior to assembly being (240 in.). Bar B is aluminum with an E of 10,000,000 psi, a C_t of 0.000 013 1 in./in./°F. Its length prior to assembly being 239 in. The assembly is to be accomplished by heating the aluminum bar until its hole center distance is 240 in. Determine (a) how many degrees the temperature of the aluminum must be increased and (b) the final distance between the centers of the steel pins after the temperature of the bars has equalized.

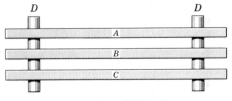

PROBLEM 3-20

3-20M *In the figure you are looking at the edges of the three bars, A, B, and C. The bars have a uniform cross section of 0.001 m². The steel pins D are*

25 mm in diameter. The bars A and C are copper with an E of 124 GPa and with the distance between the hole centers prior to assembly being 6.00 m. Bar B is aluminum with an E of 69 GPa, a C_t of 0.000 023 6 cm/cm/°C. Its length between hole centers prior to assembly was 5.975 m. The assembly is to be accomplished by heating the aluminum bar until its hole center distance is 6.00 m. Determine (a) how many degrees the temperature of the aluminum must be increased and (b) the final distance between the steel pins after the temperature of the bars has equalized.

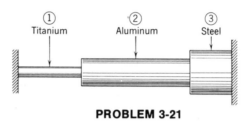

PROBLEM 3-21

3-21C   The figure shows three round bars mounted between two immovable barriers. The titanium bar has a cross-sectional area of 1 in.², a length of 4 in., an E of 15,000,000 psi, and a C_t of 0.000,003,9 in./in./°F. The aluminum bar has a cross-sectional area of 2 in.², a length of 6 in., an E of 10,000,000 psi, and a C_t of 0.000,013,1 in./in./°F. The steel bar has a cross-sectional area of 4 in.², a length of 2 in., an E of 29,000,000 psi, and a C_t of 0.000,007 in./in./°F. Determine the final lengths and the stresses developed in each of the three bars if the temperature of the system is increased 80°F.

3-21M   *The figure shows three round bars mounted between two immovable barriers. The titanium bar has a cross-sectional area of 0.0006 m², a length of 0.100 m, an E of 100 GPa, and a C_t of 0.000 007 0 m/m/°C. The aluminum bar has a cross-sectional area of 0.0012 m², a length of 0.150 m, an E of 69 GPa, and a C_t of 0.000 023 6 m/m/°C. The steel bar has a cross-sectional area of 0.0025 m², a length of 0.050 m, an E of 200 GPa, and a C_t of 0.000 0130 m/m/°C. Determine the final lengths of and the stresses developed in each of the three bars if the temperature of the system is increased 45°C.*

3-22C   The tank in the figure weighs 80,000 lb. It is supported near each end by the beam and rod system shown in the figure. The distance h_1 is 6 ft, h_2 is 10 ft, and x is 8 ft. Assume that the platform is rigid and does not bend, and that its weight is relatively insignificant. E for the steel rods is 29×10^6 psi and their F_a is 20,000 psi. Determine the smallest diameter in multiples of 1/32-in. of each rod such that f_A is not exceeded and that the platform be level under the full load of the tank.

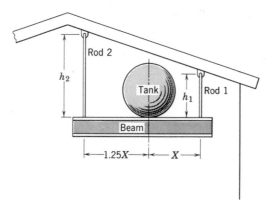

PROBLEM 3-22

3-22M *The tank in the figure weighs 35 000 kg. It is supported near each end by the beam and rod system shown in the figure. The distance h_1 is 2 m, h_2 is 3 m, and x is 2 m. Assume that the platform is rigid and does not bend. E for the steel rods is 200 GPa and their f_A is 140 MPa. Determine the smallest diameter in multiples of 5 mm of each rod such that F_a is not exceeded and that the platform be level under the full load of the tank.*

strain energy problems

3-23C A 3 in. diameter shaft 3 ft long has an axial load of 50,000 lb placed on it. Determine the strain energy in the shaft under that load. The shaft is steel with a yield strength of 42,000 lb and an E of 29,000,000 lb.

3-23M *A 75 mm shaft 1 m long has an axial load of 25 000 kg placed on it. Determine the strain energy in the shaft under that load. The shaft is steel with a yield strength of 290 MPa and an E of 200 000 MPa.*

3-24C   An aluminum tube has an outside diameter of 5.00 in. and a wall thickness of 0.500 in. Its yield strength is 18,000 psi with $E = 10,000,000$ psi. Determine its strain energy under an axial load of 100,000 lb if it is 4 ft long.

3-24M   *An aluminum tube has an outside diameter of 130 mm and a wall thickness of 15 mm. Its yield strength is 124 MPa with E = 69 GPa. Determine its strain energy under an axial load of 400 000 N if it is 1 m long.*

3-25C   Determine the energy stored in objects A and B if L is 9 in., a is 2 in., b is 3 in., and t is 1 in. Both are bronze with a yield strength of 35,000 psi and an E of 14,000,000 psi. P is 25,000 lb and it is assumed that the objects won't buckle.

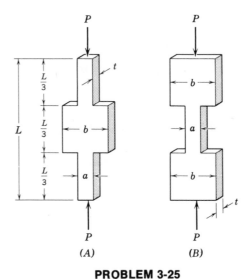

PROBLEM 3-25

3-25M   *Determine the energy stored in objects A and B if L is 30 cm, a is 5 cm, b is 8 cm, t is 3 cm, and P is 100 000 N. Both are bronze with a yield strength of 240 MPa and an E of 96 GPa.*

3-26C   Compare the resiliency of steel having a yield strength of 36,000 psi and an E of 29,000,000 psi with the resiliency of aluminum having a yield strength of 60,000 psi and an E of 10,500,000 psi.

3-26M   *Compare the resiliency of steel having a yield strength of 250 MPa and an E of 200 GPa with the resiliency of aluminum having a yield strength of 410 MPa and an E of 72 GPa.*

3-27C   How much energy can be stored in a tubular column 8 ft long having an outside diameter of 8 in. and a wall thickness of 1 in. if the modulus of resilience of the material is 94 in.-lb per in.^{3}?

3-27M   *How much energy can be stored in a tubular column 2.5 m long having an outside diameter of 200 mm and a wall thickness of 25 mm if the modulus of resilience of the material is 650 000 J/m^3?*

impact problems

3-28C   Determine the stress developed in the rod in Figure 3-8 if the weight is 20 lb, h is 10 in., L is 30 in., the rod is 0.25 in. in diameter, and the material is steel with E 29,000,000 psi.

3-28M   *Determine the stress developed in the rod in Figure 3-8 if the weight is 10 kg, h is 25 cm, L is 75 cm, the rod is 10 mm in diameter, and the material is steel with E 200 GPa.*

3-29C   In Problem 3-28C determine the largest value h could have and still not cause any stress greater than 36,000 psi in the rod.

3-29M   *In Problem 3-28M determine the largest value h could have and still not cause any stress greater than 250 MPa in the rod.*

3-30C   A weight is dropped on the end of a vertical shaft 3 in. in diameter and 24 in. long causing it to deform 0.0015 in. If the weight is dropped from a point 20 in. above the end of the shaft which has an E of 29,000,000 psi, what is the magnitude of the weight?

3-30M   *A weight is dropped on the end of a vertical shaft 75 mm in diameter and 60 cm long causing it to deform 0.0040 mm. If the weight is dropped from a point 50 cm above the end of the shaft, which has an E of 200 GPa, what is the magnitude of the weight?*

3-31C   A spring is designed to absorb the energy of a 3000 lb automobile traveling at 5 mph while deflecting no more than 3 in. Determine how much force must be required to deflect the spring each inch of its deflection.

3-31M   *A spring is designed to absorb the energy of a 1500 kg automobile traveling at 8 km/h without deflecting the spring more than 0.08 m. Determine the force that must be required to deflect the spring each centimeter of its deflection*

3-32C   A 1 oz bullet traveling at 2000 fps impacts on the end of a 1 in. square steel bar. The end of the rod has an impervious surface so that the bullet causes no penetration. Determine the maximum stress and deflection of the steel bar due to the impact of the bullet. The bar is 10 in. long and has an *E* of 29,000,000 psi.

3-32M   *A 0.03 kg bullet traveling at 600 m/s impacts on the end of a 3 cm square bar. The end of the bar has an impervious surface so that the bullet causes no penetration. Determine the maximum stress and deflection of the steel bar due to the impact of the bullet. The bar is 0.25 m long and has an E of 200 GPa.*

CHAPTER
OBJECTIVES ⎯⎯⎯⎯⎯⎯⎯⎯⎯⎯⎯⎯⎯⎯

The purpose of your study of this chapter is to

1. Learn what a pressure vessel is.

2. Learn the difference between pressure vessels classified as thin walled and those classified as thick walled.

3. Understand the nature of the forces in pressure vessels.

4. Learn how to determine the stresses in the walls of thin-walled cylinders.

5. Learn how to determine the stresses in the walls of thin-walled, spherical pressure vessels.

CHAPTER ——————————— 4
THIN-WALLED PRESSURE VESSELS

Pressure vessels are containers that hold fluids under pressure. Examples are tanks, boilers, pipe lines, and pressure cookers. We are all familiar with some of these. We see tank trucks on the highways that contain all kinds of fluids, some of which are gases under relatively high pressures. We fill our automobile tires with air at the gas station—it comes from a storage tank of air maintained at high pressure. Your automobile tires themselves are pressure vessels. Failure of pressure vessels can cause loss of life either by catastrophic failure, by explosion, or by simple failure such as leakage permitting lethal or explosive gases to escape in the vicinity of people. Therefore the design of vessels containing fluids under high pressures is carefully controlled by regulatory codes such as the ASME Boiler Code and the ASME Pressure Vessel Code.

4-1 DIFFERENCE BETWEEN THIN-WALLED AND THICK-WALLED PRESSURE VESSELS

The walls of pressure vessels are subjected to three-dimensional stress. Up to a point the designer can assume that the stress is evenly distributed through the wall. The point up to which this assumption is considered safe is where the wall thickness is no more than one-tenth the inside diameter of the vessel. Thus our concern will be with pressure vessels having an inside diameter that is at least ten times the thickness of the wall. Also, we will assume that all of our vessels are welded at their seams, as is the current practice, and that the welds are as strong as the base metal of the vessels. Thus we need not be concerned with joint strengths and will in effect consider the vessels to be jointless.

4-2 FORCES IN PRESSURE VESSELS

Fluids (liquid or gaseous) under pressure exert equal forces in all directions. *Pressure* is the force distributed over an area, and is expressed in the customary system as pounds per square inch (psi) and in the metric system as Newtons per square meter (Pa).

Figure 4-1 shows a pressure cylinder having spherical ends. Through the

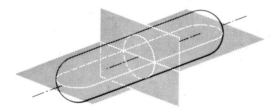

FIGURE 4-1

cylinder cut two planes, one perpendicular to the longitudinal axis, the other passing through the longitudinal axis. In the succeeding paragraphs we will analyze the forces that act on the surfaces cut by these two principal planes. In the case of a sphere the surface cut by any plane passing through the center of the sphere is all that need be analyzed.

In Figure 4-2 is shown the cross section of a cylinder or sphere. The small arrows indicate the force being exerted by the pressurized fluid; each arrow represents an increment of that pressure. Each arrow acts perpendicular to

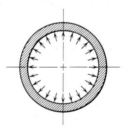

FIGURE 4-2

the surface of the section. Each arrow is matched by one equal and opposite to it on the opposite side of the section. This is an accurate model of the forces acting on the section and from it the deduction can be made that there is no shearing force caused by the pressure. Therefore in our analysis we need only to consider the tensile forces.

4-3   THIN-WALLED CYLINDERS

We will first analyze the cross section formed by passing a plane vertically through the longitudinal centerline of a cylinder as shown in Figure 4-3. We will call this cross section the *longitudinal section.* Note that the ends of the

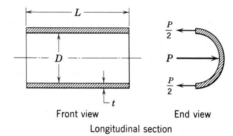

Front view   End view
Longitudinal section

FIGURE 4-3

cylinder are not shown because we are examining the straight portion of the cylinder with the knowledge that the pressure is being maintained therein. In the end view we see a free-body diagram of the section with the force P representing the sum of all of the components of pressure acting in the horizontal direction. By summing forces in the horizontal direction we find that the total force acting on each of the cut edges must be $P/2$ if the section is to be in static equilibrium. In the front view of the section the area held in tension by each of the forces $P/2$ is shown by cross-hatching and can be seen to be L times t each. The stress in the longitudinal section is

$$f_L = \frac{F}{A} = \frac{P/2}{Lt} = \frac{P}{2Lt}$$

It can be determined mathematically that

$$P = pLD$$

where p is the pressure in psi or PA and D is the inside diameter of the cylinder.

Substituting this in the above equation yields

$$f_L = \frac{pLD}{2Lt} = \frac{pD}{2t} \tag{4-1}$$

where f_L is the stress in the longitudinal section, p the pressure, D the inside diameter of the cylinder, and t the thickness of the cylinder.

Example

customary

Determine the stress in the longitudinal section of a cylinder 60 in. in diameter, 1 in. thick, which contains gas at a pressure of 150 psi.

metric

Determine the stress in the longitudinal section of a cylinder 1.5 m in diameter, 25 mm thick, which contains gas at a pressure of 1 MPa.

Solution

customary

$$f_L = \frac{pD}{2t} = \frac{150 \text{ psi} (60 \text{ in.})}{2\,(1 \text{ in.})}$$
$$= 4500 \text{ psi}$$

metric

$$f_L = \frac{pD}{2t} = \frac{1 \text{ MPa} (1.5 \text{ m})}{2\,(0.025 \text{ m})} = 30 \text{ MPa}$$

We will call the cross section of a cylinder cut by a plane perpendicular to its longitudinal axis its *circumferential section*. Referring to Figure 4-4 we see that the force P tends to separate the cylinder in the longitudinal direction. P

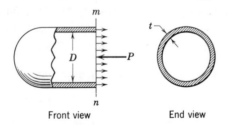

Front view   End view

FIGURE 4-4

is the force of the pressure acting over the cross-sectional area having an inside diameter D.

$$P = pA = p\frac{\pi D^2}{4}$$

The area of the metal cut is very close to $\pi D_m t$, where D_m is the mean diameter of the cylinder wall. Therefore, the stress in the circumferential section is for practical purposes

$$f_{ci} = \frac{p(\pi D^2/4)}{\pi D_m t} = \frac{pD^2}{4D_m t} \tag{4-2}$$

Example

customary

Determine the stress in a circumferential section of a cylinder having a diameter of 36 in. and a wall thickness of 0.5 in. when filled with a gas at 625 psi pressure.

metric

Determine the stress in a circumferential section of a cylinder having a diameter of 1 m and a wall thickness of 10 mm when filled with a gas at 4 MPa.

Solution

customary

$$f_{ci} = \frac{pD^2}{4D_m t} = \frac{625 \text{ psi } (36 \text{ in.})^2}{4 (36 \text{ in.} + 0.5 \text{ in.})(0.5 \text{ in.})}$$
$$= 11,100 \text{ psi}$$

metric

$$f_{ci} = \frac{pD^2}{4D_m t} = \frac{4 \text{ MPa } (1 \text{ m})^2}{4 (1 \text{ m} + 0.01 \text{ m})(0.01 \text{ m})}$$
$$= 99 \text{ MPa}$$

The stress in the longitudinal section is generally called the circumferential stress because it acts in the circumferential direction, while the stress in the circumferential section is generally called the longitudinal stress because it acts in the longitudinal direction. We will not use those terms here because they tend to be somewhat confusing.

4-4 THIN-WALLED SPHERICAL PRESSURE VESSELS

A *thin-walled spherical pressure vessel* like a thin-walled cylinder is one in which the inside diameter is at least 10 times the wall thickness. The maximum stress is developed in that section of the wall which is cut by a plane passing through the center of the sphere. The force tending to separate the vessel on this cutting plane is

$$P = p\frac{\pi D^2}{4}$$

The stress is equal to the above force P divided by the area of the cut section which is $\pi D_m t$, where D_m is the mean diameter of the spherical shell. This gives us

$$f_s = \frac{p(\pi D^2/4)}{\pi D_m t} = \frac{pD^2}{4D_m t} \tag{4-3}$$

where f_s is the maximum stress in the spherical pressure vessel.

Example

customary	metric
Determine the maximum stress developed in the wall of a spherical pressure vessel having an inside diameter of 16 ft and a wall thickness of 1 in., and which contains liquified propane gas at a pressure of 600 psi.	*Determine the maximum stress developed in the wall of a spherical pressure vessel having an inside diameter of 5 m and a wall thickness of 25 mm, and which contains liquified propane gas at a pressure of 4.1 MPa.*

Solution

customary

$$f_s = \frac{pD^2}{4D_m t}$$

$$= \frac{(600\ \text{psi})(16\ \text{ft} \times 12\ \text{in./ft})^2}{4\,(16\ \text{ft} \times 12\ \text{in./ft} + 1\ \text{in.})(1\ \text{in.})}$$

$$= 28{,}700\ \text{psi}$$

metric

$$f_s = \frac{pD^2}{4D_m t}$$

$$= \frac{4.1\ \text{MPa}\ (5\ \text{m})^2}{4\,(5\ \text{m} + 0.025\ \text{m})(0.025\ \text{m})}$$

$$= 204\ \text{MPa}$$

4-5   ACCOMPLISHMENT CHECKLIST

From your studies of this chapter do you now

1.   Understand what a pressure vessel is.

2.   Know the difference between a thin-walled and a thick-walled pressure vessel.

3.   Understand the nature of forces in pressure vessels and why no shear stresses are caused by those forces.

4.   Know how to determine the stresses in or the required thicknesses of the walls of thin-walled cylinders.

5.   Know how to determine the stresses in or the required thicknesses of the walls of thin-walled spherical pressure vessels?

4-6   SUMMARY

Pressure vessels are containers that hold fluids under pressure. A *thin-walled pressure vessel* is one having an inside diameter that is at least 10 times the wall thickness. *Pressure* is a force distributed over an area. The *longitudinal section* of a cylinder is that area cut by a plane which is parallel to and passes through the longitudinal axis. The *circumferential section* of a cylinder is that area cut by a plane passing through the cylinder perpendicular to its longitudinal axis. The stress in the longitudinal section of a cylinder is

$$f_L = \frac{pD}{2t} \qquad\qquad (4\text{-}1)$$

where f_L is the stress in the longitudinal section, p is the pressure in psi or Pa, D is the inside diameter of the cylinder, and t is its wall thickness.

The stress in a circumferential section of a cylinder is

$$f_{ci} = \frac{pD^2}{4D_m t} \qquad\qquad (4\text{-}2)$$

where f_c is the stress in the circumferential section of a cylinder. The

91

maximum stress developed in the wall of a spherical pressure vessel is

$$f_s = \frac{pD^2}{4D_m t}$$ (4-3)

where f_s is the maximum stress developed in the wall of a spherical pressure vessel and the other symbols stand for the same things as in the preceding equations for cylinders.

PROBLEMS

4-1C   Determine the stresses in the longitudinal and circumferential sections of a cylinder with spherical ends containing air at 250 psi pressure. The cylinder is 3 ft in diameter and has a wall thickness of 0.25 in.

4-1M   *Determine the stresses in the longitudinal and circumferential sections of a cylinder with spherical ends containing air at a pressure of 2 MPa. The cylinder is 1 m in diameter and has a wall thickness of 8 mm.*

4-2C   Determine the maximum pressure that can be contained by a cylinder having a diameter of 30 in. and a wall thickness of 0.375 in. without exceeding the allowable stress of 15,000 psi.

4-2M   *Determine the maximum pressure that can be contained by a cylinder having a diameter of 75 cm and a wall thickness of 10 mm without exceeding the allowable stress of 100 MPa.*

4-3C   Determine the minimum wall thickness required for a cylinder 4 ft in diameter containing gas at a pressure of 600 psi if the allowable stress for the material is 14,000 psi.

4-3M   *Determine the minimum wall thickness required for a cylinder 1.25 m in diameter containing gas at a pressure of 4 MPa if the allowable stress for the material is 95 MPa.*

4-4C   The steam drum (a horizontal cylinder) of a power plant when half full of water is producing steam at a pressure of 960 psi. Determine the minimum wall thickness for the drum if it is 48 in. in diameter and is made of ASTM A 515, Grade 55 steel with a minimum yield stress of 30,000 psi.

4-4M   *The steam drum (a horizontal cylinder) of a power plant when half full of water is producing steam at a pressure of 6 MPa. Determine the minimum wall thickness for the drum if it is 1.2 m in diameter and is made of ASTM A 515, Grade 55 steel with a minimum yield stress of 200 MPa.*

4-5C   A standpipe filled with water is 100 ft high. Its inside diameter is 6 in. and its wall thickness is 0.125 in. Determine the stress in the pipe due to the water pressure. Water weighs 62.4 lb/ft^3.

4-5M   *A standpipe filled with water is 30 m high. Its inside diameter is 15 cm and its wall thickness is 3 mm. Determine the stress in the pipe due to the water pressure. Water weighs 1000 kg/m^3.*

4-6C   Tubing is generally specified by its outside diameter. At what pressure would a 1.50 in. aluminum tube with a wall thickness of 0.125 in. burst if it were made of 3003-18 aluminum with an ultimate strength of 27,000 psi?

4-6M   *Tubing is generally specified by its outside diameter. At what pressure would a 40 mm aluminum tube with a wall thickness of 3 mm burst if it were made of 3003-18 aluminum with an ultimate strength of 180 MPa?*

4-7C   A spherical pressure vessel contains chlorine at a pressure of 1200 psi. Its diameter is 6 ft and its thickness is 6 in. Determine the stress in the wall.

4-7M   *A spherical pressure vessel contains chlorine at 8 MPa. Its diameter is 2 m and its thickness is 150 mm. Determine the stress in the wall.*

4-8C   Liquid oxygen is to be stored in a spherical tank 8 ft in diameter. It is anticipated that the pressure will go no higher than 800 psi. Determine the minimum wall thickness if the maximum allowable stress is 14,500 psi.

4-8M *Liquid oxygen is to be stored in a spherical tank 2.5 m in diameter. It is anticipated that the pressure will go no higher than 5.50 MPa. Determine the minimum wall thickness if the maximum allowable stress is 100 MPa.*

4-9C Determine the maximum allowable diameter for a cylindrical pressure vessel designed to contain air at a maximum pressure of 800 psi if the vessel is to have a wall thickness of 2 in. and is made of steel having an allowable stress of 14,500 psi.

4-9M *Determine the maximum allowable diameter for a cylindrical pressure vessel designed to contain air at a maximum pressure of 5.5 MPa if the vessel is to have a wall thickness of 50 mm and is made of steel having an allowable stress of 100 MPa.*

4-10C Determine the maximum allowable diameter for a spherical pressure vessel designed to contain air at a maximum pressure of 800 psi if the vessel is to have a wall thickness of 2 in. and is made of steel having an allowable stress of 14,500 psi.

4-10M *Determine the maximum allowable diameter for a spherical pressure vessel designed to contain air at a maximum pressure of 5.5 MPa if the vessel is to have a wall thickness of 50 mm and is made of steel having an allowable stress of 100 MPa.*

4-11C Determine the maximum stress in the system shown in the figure if the standpipe is 12 in. in diameter, the cylinder is 60 in. in diameter, the wall

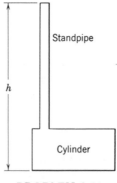

PROBLEM 4-11

thickness of the standpipe and cylinder is 3/16 in., and h is 200 ft. Water weighs 62.4 lb/ft³.

4-11M *Determine the maximum stress in the system shown in the figure if the standpipe is 300 mm in diameter, the cylinder is 1.50 m in diameter, the wall thickness of the standpipe and cylinder is 5 mm, and h is 60 m. Water weighs 1 000 kg/m³.*

4-12C In the figure is a system consisting of a standpipe having a 12 in. diameter, a cylinder with flat ends with a diameter of 36 in., and a sphere of 60 in. diameter. The wall thickness of all members is 0.25 in. The material is steel with an allowable stress of 15,000 psi. Determine the maximum height to which the standpipe could be filled with water (62.4 lb/ft³) without exceeding the allowable stress in any member assuming that the connections between members will not fail.

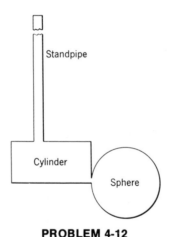

PROBLEM 4-12

4-12M *In the figure is a system consisting of a standpipe having a 0.300 m diameter, a cylinder with flat ends with a diameter of 0.900 m, and a sphere with a diameter of 1.50 m. The wall thickness of all members is 5 mm. The material is steel with an allowable stress of 100 MPa. Determine the maximum height to which the standpipe could be filled with water (1 000 kg/m³) without exceeding the allowable stress in any member assuming that the connections between members will not fail.*

95

CHAPTER
OBJECTIVES

The object of studying this chapter is to

1. Learn of some of the different types of connections.

2. Become familiar with the terminology of bolted and riveted joints.

3. Learn the modes of failure of bolted and riveted joints.

4. Learn how to analyze the various types of bolted and riveted joints.

5. Learn how to analyze eccentrically loaded joints.

6. Become familiar with some of the more important types of welded joints and the related terminology.

7. Learn how and why welded joints may fail.

8. Learn how to design joints using fillet and butt welds.

9. Learn how to determine the strengths of resistance welds.

CHAPTER ———————————— 5
CONNECTIONS

There are many methods of fastening members of structures and machines together; welding, bolting, riveting, stapling, adhesive bonding, and many variations of each. We will limit our study to structural connections made by bolting, riveting, and welding. Until recently, such a study would have been limited to one material—steel. Now, other materials are gaining importance, notably aluminum with its extensive use in air craft and transportation equipment. The basic theory is the same for the analysis of joints of different materials. There are, however, some differences in requirements and procedures which we will examine and use. Design requirements for steel structures are regulated by the American Steel Institute (ASI), the American Institute of Steel Construction (AISC), and the American Association of State Highway Officials (AASHO), among others. Design requirements for aluminum structures are largely controlled by the Aluminum Association.

5-1   TYPES OF BOLTED AND RIVETED CONNECTIONS

The two simplest joints are the butt joint and the lap joint. In the *lap joint* the pieces to be connected are simply lapped over each other and bolted or riveted as shown in Figure 5-1. In the *butt joint* the main plates are brought together end to end and are then fastened together using *cover plates* as shown in Figure 5-1. These joints may be further classified as continuous or structural. A *continuous joint* is used to join two relatively thin sheets of material, such as the aluminum skin sections of aircraft where the joint

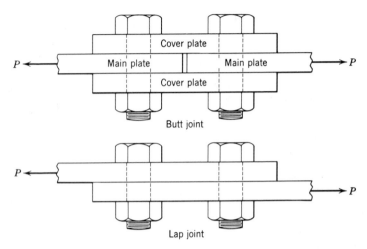

FIGURE 5-1

amounts to a long seam which may involve the use of hundreds of rivets (rivets being the common connectors used for this type of joint).

A *structural joint* is used to connect structural members such as bars, plates, angles, channels, beams, and columns. It may be a simple angle to plate connection as in Figure 5-2, or one of the more complex framed or

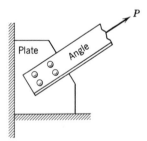

FIGURE 5-2

seated connections. *Framed connections* are those such as shown in Figure 5-3, which shows one beam suspended from another by the use of a pair of angles. A *seated connection* is one in which the member actually sits on a seat formed by an angle as shown in Figure 5-4.

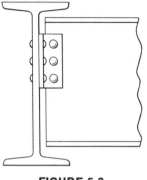

FIGURE 5-3

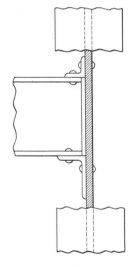

FIGURE 5-4

5-2   FAILURE OF BOLTED AND RIVETED JOINTS

Bolted and riveted joints are essentially the same except for the fasteners (bolts or rivets). Although there are some very important differences between the effects of using these fasteners, the manner of failure is basically the same. Since the use of rivets has virtually ceased in steel construction, our analyses of riveted joints will be confined to aluminum construction with

99

aluminum rivets, using the design conventions in connection with that material. Material, examples, and problems dealing with bolted joints will be applicable to steel or aluminum because the calculations would be the same for either.

The three ways in which riveted and bolted joints fail are shear failure, bearing failure, or failure by tearing (tensile failure) of the joined members or cover plates. *Shear failure* occurs when the fasteners are sheared off as shown in Figure 5-5(a). *Bearing failure* occurs when the material behind the fastener crushes or when the fastener itself crushes. The latter would occur only when the fastener was softer than the members being joined, which is extremely rare. Bearing failure is shown diagramatically in Figure 5-5(b). *Failure by tearing* is shown in Figure 5-5(c). In a lap joint either plate could tear, while in the case of a butt joint with cover plates the failure could occur in either the cover plates or the main plates. In the following section we will analyze joints by determining the forces and stresses which would cause them to experience these types of failures.

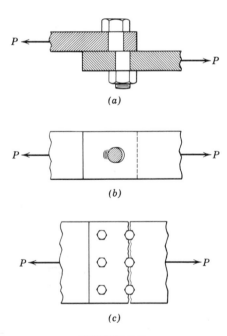

(a)

(b)

(c)

FIGURE 5-5

5-3   ANALYSIS OF LAP AND BUTT JOINTS

For a joint to be most efficient the applied load should pass through the centroid of the connector pattern and such loading (concentric loading) is assumed in this section. The analysis of joints in which the applied load does not pass through the centroid of the rivet pattern (eccentrically loaded joints) is deferred to section 5-4. As concentrically loaded joints are assembled the bolts or rivets may initially take unequal shares of the load. As the load is increased, however, the elastic properties of the materials cause the members to adjust so that the load becomes (for all practical purposes) evenly divided.

The basic procedures in analyzing bolted and riveted joints can be demonstrated by first analyzing a simple lap joint and then a butt joint with cover plates.

Analysis of concentrically loaded lap joint

We will first analyze the joint for shear of the bolts. Because the load is evenly divided between the bolts, the shear stress in the bolts can be determined by

$$f_v = \frac{P}{NA} \tag{5-1}$$

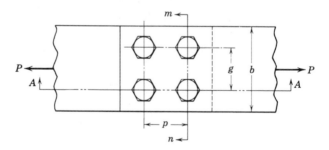

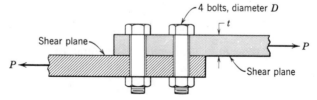

FIGURE 5-6 Section A-A.

where f_v is the shear stress in the connector, P is the total load on the joint, N is the number of connectors, and A is the cross-sectional area of each bolt.

In the case of an aluminum riveted structure the shear stress should be figured on the basis of the cross-sectional area of the hole (see Appendix 6) since the rivets are required to fill the hole.

In determining the bearing stress, the area supporting the bearing load in the case of bolts is the entire cross section of the material immediately behind the bolts. For aluminum rivets the bearing area is the entire cross section behind the rivet holes. Thus

$$f_c = \frac{P}{NtD} \tag{5-2}$$

where f_c is the compressive (bearing) stress in the material behind the connectors, P is the total load on the joint, N is the number of connectors, t is the thickness of the member being analyzed (if members of the same material but of more than one thickness are being connected, the calculation should be made for the thinner), and D is the diameter of the bolt or the diameter of the hole for the aluminum rivet.

Finally, examination of Figure 5-6 shows us that if the plate were over-stressed in tension, the upper plate could fail at section mn. The total area of the plate's cross section (without holes) is the *gross area*. The *net area* is the gross area minus the area of the cross sections of the holes. It is

$$A_n = bt - N_s D_h t = (b - N_s D_h)t \tag{5-3a}$$

where A_n is the net area, b is the width of the section, N_s is the number of holes in the section mn, t is the thickness of the upper plate (if members of different thicknesses of the same material are being connected, only a calculation of the thinner member need be made), and D_h is the diameter of the holes. D_h is determined as follows.

For aluminum rivets see Appendix 6.
For bolts

	Customary Units	Metric Units
Drilled holes	$D + 1/32$ in.	$D + 0.0008$ m
Punched holes	$D + 1/8$ in.	$D + 0.0032$ m

where D is the nominal bolt diameter.

The stress in section mn can be found by

$$f_t = \frac{P}{A_n} = \frac{P}{(b - N_s D_h)t} \tag{5-3}$$

where f_t is the tensile stress on the net cross section, P is the total load on the cross section, and A_n is the net area determined by Equation 5-3a.

Example

customary

Determine the maximum shear, bearing, and tensile stresses in the lap joint in Figure 5-6 if $b = 4$ in., $t = 3/8$ in., $g = 2$ in., $p = 1\text{-}1/2$ in., $D = 3/4$ in., and $P = 20{,}000$ lb (20 kips). Holes are punched.

metric

Determine the maximum shear, bearing, and tensile stresses in the lap joint in Figure 5-6 if $b = 100$ mm, $t = 10$ mm, $p = 40$ mm, $D = 20$ mm, and $P = 100\,000$ N. Holes are punched.

Solution

customary

Find the maximum shear stress in the bolts using Equation 5-1.

$$A = \frac{\pi D^2}{4} = 0.442 \text{ in.}^2$$

$$f_v = \frac{P}{NA} = \frac{20{,}000 \text{ lb}}{4\,(0.442 \text{ in.}^2)}$$
$$= 11{,}300 \text{ psi}$$

Next we find the maximum bearing stress using Equation 5-2.

$$f_c = \frac{P}{NtD} = \frac{20{,}000 \text{ lb}}{4\,(0.375 \text{ in.})(0.750 \text{ in.})}$$
$$= 17{,}800 \text{ psi}$$

The maximum tensile stress in section *mn* is found by using Equations 5-3a and 5-3.

$$D_h = D + 1/8 \text{ in.} = 0.75 \text{ in.} + 0.125 \text{ in.}$$
$$= 0.875 \text{ in.}$$

$$A_n = (b - N_s D_h)t$$
$$= (4 \text{ in.} - 2(0.875 \text{ in.}))0.375 = 0.844 \text{ in.}^2$$

$$f_t = \frac{P}{A_n} = \frac{20{,}000 \text{ lb}}{0.844 \text{ in.}^2} = 23{,}700 \text{ psi}$$

metric

Find the maximum shear stress in the bolts using Equation 5-1.

$$A = \frac{\pi D^2}{4} = \frac{\pi\,(0.020 \text{ m})^2}{4} = 3.14 \times 10^{-4} \, m^2$$

$$f_v = \frac{100\,000 \text{ N}}{(4)(3.14 \times 10^{-4} \, m^2)} = 79.6 \text{ MPa}$$

Next we find the maximum bearing stress using Equation 5-2.

$$f_c = \frac{P}{NtD} = \frac{100\,000 \text{ N}}{(4)(0.010 \text{ m})(0.020 \text{ m})}$$
$$= 125 \text{ MPa}$$

The maximum tensile stress in section mn is found by using Equations 5-3a and 5-3.

$$D_h = D + 0.0032\,m = 0.020\,m + 0.0032\,m$$
$$= 0.0232\,m$$

$$A_n = (b - N_s D_h)t$$
$$= (0.100\,m - 2(0.0232\,m))0.010\,m$$
$$= 5.36 \times 10^{-4} \, m^2$$

$$f_t = \frac{P}{A_n} = \frac{100\,000 \text{ N}}{5.36 \times 10^{-4} \, m^2} = 187 \text{ MPa}$$

Analysis of butt joint

Figure 5-7 shows a butt joint with a bolt pattern of five bolts on each side of the joint. It should be noted that a small space is left between the main plates. In section A-A the joint is shown in the loaded condition with the left-hand main plate shifted toward the right bringing pressure to bear on the left-hand side of the bolts. The cover plates resist this pressure by exerting a pressure on the right-hand side of the bolts. A free-body diagram of a bolt would appear as in Figure 5-8. It can be assumed that the load is equally divided among the five bolts. In the case of the butt joint there are two areas of the bolt in shear, that on plane ab and that on plane cd in the figure. Therefore, the bolt is said to be in *double shear* and the shear stress in the bolt is

$$f_{vb} = \frac{P}{2NA} \qquad (5\text{-}4)$$

where f_{vb} is the shearing stress in a butt joint, P is the total load on the joint, N is the number of connectors on one side of the joint, and A is the cross-sectional area of each connector.

Just as each bolt assumes its share of the total load, it also transmits that share to the cover plates. Because there are two cover plates, the bolts exert only half as much force on each of the cover plates as on the main plate. Consequently, it is necessary to make separate calculations for the main and cover plates.

The main plate bearing stress is

$$f_{cm} = \frac{P}{NDt} \qquad (5\text{-}5)$$

The cover plate bearing stress is

$$f_{cc} = \frac{P}{2NDt} \qquad (5\text{-}6)$$

When determining the tensile stress in the joint we must not only consider the main and cover plates separately, but must also consider cross sections through each row of connectors in the main and cover plates separately. With experience it may sometimes be possible to determine the critical sections by examination. If there is any doubt, however, *calculate*.

Examining the main plate of the joint in Figure 5-7 we can see that the force to the right of Row 1 is the total load P, which is the force tending to tear the plate across Row 1. The stress can be determined by using Equation 5-3 with $N_s = 2$.

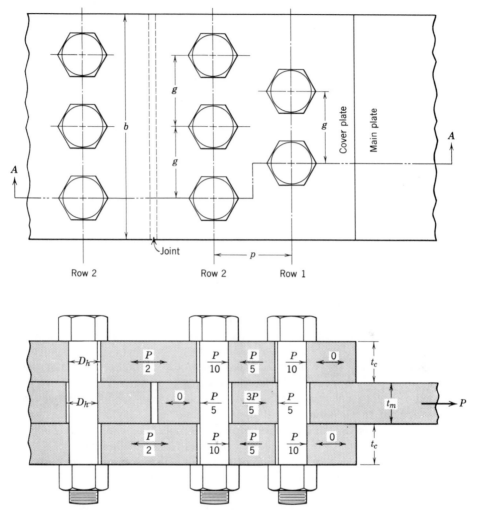

FIGURE 5-7   Section A-A. *Note*: forces in bolts are for the individual bolt only, forces in plates are for the entire cross-section.

As Row 1 of fasteners is passed they transfer their portion of the load $2P/5$ to the cover plates, leaving $3P/5$ in the main plate. Thus $3P/5$ is the force tending to tear the main plate at Row 2. The stress in the section at Row 2 may be determined by again using Equation 5-3 with $3P/5$ being used in place of P.

Examining the cover plates it is seen that the two bolts in Row 1 transmit a $P/5$ force to each of the two cover plates. It is this force that tends to tear the

105

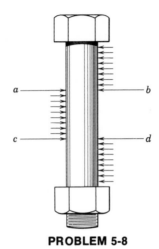

PROBLEM 5-8

cover plates at Row 1. The stress in the cover plates at Row 1 is found by using Equation 5-3, by using $P/5$ in place of P.

The three bolts in Row 2 transfer the balance of the load P to the cover plates. Each cover plate now has a $P/2$ force in it tending to tear the cover plates across Row 2. Again we apply Equation 5-3 substituting $P/2$ for P.

Thus we can determine the stresses in all members of the joint in all the critical places. There are other possible modes of failure such as crushing or bending the connectors or tearing the plate diagonally. Generally it is not necessary to check these because standard joint design practices prevent failure in these modes. If, it does become necessary, the extension of the principles used in the foregoing analyses will take care of determining the bearing stresses in the connectors or the stresses in the diagonal sections of the connector pattern. Bending of the connectors sometimes occurs in joints having several thicknesses of plates, causing the connectors to be quite long in comparison with their diameter. The determination of the bending of connectors in such situations is beyond the scope of this text.

Example

customary

Determine the shear, bearing, and tensile stresses in the joint in Figure 5-7 if $D =$ 0.875 in., $b = 5.000$ in., $t_c = 0.375$ in., $t_m =$

metric

Determine the shear, bearing, and tensile stresses in the joint in Figure 5-7 if $D = 24$ mm, $b = 125$ mm, $t_c =$

0.625 in., and $P = 30{,}000$ lb. Holes are punched. $D_h = D + 0.125$ in. $= 1.00$ in.

$10\,mm$, $t_m = 16\,mm$, and $P = 130\,kN$. Holes are punched. $D_h = D + 0.0032\,m = 0.0272\,m$.

Solution

customary

Use Equation 5-4 to find the shearing stress in the bolts.

$$f_{vb} = \frac{P}{2NA} = \frac{30{,}000 \text{ lb}}{2\,(5)(0.601 \text{ in.}^2)} = 4990 \text{ psi}$$

Use Equation 5-5 to find the bearing stress in the main plate.

$$f_{cm} = \frac{P}{NDt_m} = \frac{30{,}000 \text{ lb}}{5\,(0.875 \text{ in.})(0.625 \text{ in.})}$$
$$= 11{,}000 \text{ psi}$$

Use Equation 5-6 to find the bearing stress in the cover plates.

$$f_{cc} = \frac{P}{2NDt_c} = \frac{30{,}000 \text{ lb}}{2\,(5)(0.875 \text{ in.})(0.375 \text{ in.})}$$
$$= 9140 \text{ psi}$$

Determine the stress in a section through Row 1 of the main plate by use of Equations 5-3a and 5-3.

$$A_n = (b - N_s D_h)t_m = (5.00 \text{ in.} - 2(1 \text{ in.}))$$
$$\times 0.625 \text{ in.} = 1.875 \text{ in.}^2$$

$$f_t = \frac{P}{A_n} = \frac{30{,}000 \text{ lb}}{1.875 \text{ in.}^2} = 16{,}000 \text{ psi}$$

Determine the stress in a section through Row 2 of the main plate by use of Equations 5-3a and 5-3. The total load on this section is $3P/5$ and N_s is 3.

metric

Use Equation 5-4 to find the shearing stress in the bolts.

$$f_{vb} = \frac{P}{2NA} = \frac{130 \text{ kN}}{2\,(5)(4.52 \times 10^{-4})}$$
$$= 28.8 \text{ MPa}$$

Use Equation 5-5 to find the bearing stress in the main plate.

$$f_{cm} = \frac{P}{NDt_m} = \frac{130 \text{ kN}}{5\,(0.024 \text{ m})(0.016 \text{ m})}$$
$$= 67.7 \text{ MPa}$$

Use Equation 5-6 to find the bearing stress in the cover plates.

$$f_{cc} = \frac{P}{2NDt_c} = \frac{130 \text{ kN}}{2\,(5)(0.024 \text{ m})(0.010 \text{ m})}$$
$$= 54.2 \text{ MPa}$$

Determine the stress in a section through Row 1 of the main plate by use of Equations 5-3a and 5-3.

$$A_n = (b - N_s D_h)t_m$$
$$= (0.125 \text{ m} - 2(0.0272 \text{ m}))$$
$$\times 0.016 \text{ m} = 0.001\,13 \text{ m}^2$$

$$f_t = \frac{P}{A_n} = \frac{130 \text{ kN}}{0.001\,13 \text{ m}^2} = 115 \text{ MPa}$$

Determine the tensile stress in a section through Row 2 of the main plate by use of Equations 5-3a and 5-3. The total load on this section is $3P/5$ and N_s is 3.

$$A_n = (b - N_sD_h)t_m = (5.00 \text{ in.} - 3(1 \text{ in.}))$$
$$\times 0.625 \text{ in.} = 1.25 \text{ in.}^2$$
$$f_t = \frac{3P/5}{A_n} = \frac{3P}{5A_n} = \frac{3(30,000 \text{ lb})}{5(1.25 \text{ in.}^2)}$$
$$= 14,400 \text{ psi}$$

Determine the tensile stress in a section through Row 1 of the cover plates by use of Equations 5-3a and 5-3. Note that the total load on this section is $P/5$ and N_s is 2.

$$A_{n'} = (b - N_sD_h)t_c = (5.00 \text{ in.} - 2(1 \text{ in.}))$$
$$\times 0.375 \text{ in.} = 1.125 \text{ in.}^2$$
$$f_t = \frac{P/5}{A_n} = \frac{P}{5A_n} = \frac{30,000 \text{ lb}}{5(1.125 \text{ in.}^2)}$$
$$= 5330 \text{ psi}$$

Determine the tensile stress in a section through Row 2 of the cover plates by using Equations 5-3a and 5-3. In this case the force on the section is $P/2$ and N_s is 3.

$$A_n = (b - N_sD_h)t_c = (5 \text{ in.} - 3(1 \text{ in.}))$$
$$\times 0.375 \text{ in.} = 0.750 \text{ in.}^2$$
$$f_t = \frac{P/2}{A_n} = \frac{P}{2A_n} = \frac{30,000 \text{ lb}}{2(0.750 \text{ in.}^2)}$$
$$= 20,000 \text{ psi}$$

Reviewing the stresses we find that the highest stress is in Row 2 of the cover plates. Hence, Row 2 is the most likely place for failure to occur.

$$A_n = (b - N_sD_h)t_m$$
$$= (0.125 \text{ m} - 3(0.0272 \text{ m}))$$
$$\times 0.016 \text{ m} = 6.94 \times 10^{-4} \text{ m}^2$$
$$f_t = \frac{3P/5}{A_n} = \frac{3P}{5A_n} = \frac{3(130 \text{ kN})}{5(6.94 \times 10^{-4} \text{ m}^2)}$$
$$= 112 \text{ MPa}$$

Determine the tensile stress in a section through Row 1 of the cover plates by use of Equations 5-3a and 5-3. Note that the total load on this section is P/5 and N_s is 2.

$$A_n = (b - N_sD_h)t_c$$
$$= (0.125 \text{ m} - 2(0.0272 \text{ m}))$$
$$\times 0.010 \text{ m} = 7.06 \times 10^{-4} \text{ m}^2$$
$$f_t = \frac{P/5}{A_n} = \frac{P}{5A_n} = \frac{130 \text{ kN}}{5(7.06 \times 10^{-4} \text{ m}^2)}$$
$$= 36.8 \text{ MPa}$$

Determine the tensile stress in a section through Row 2 of the cover plates by using Equations 5-3a and 5-3. In this case the force on the section is P/2 and N_s is 3.

$$A_n = (b - N_sD_h)t_c$$
$$= (0.125 \text{ m} - 3(0.0272 \text{ m}))$$
$$\times 0.010 \text{ m} = 4.34 \times 10^{-4} \text{ m}^2$$
$$f_t = \frac{P/2}{A_n} = \frac{P}{2A_n} = \frac{130 \text{ kN}}{2(4.34 \times 10^{-4} \text{ m}^2)}$$
$$= 150 \text{ MPa}$$

Reviewing the stresses we find that the highest stress is in Row 2 of the cover plates. Hence, Row 2 is the most likely place for failure to occur.

5-4 ECCENTRICALLY LOADED STRUCTURAL JOINTS

As previously mentioned, for greatest efficiency the applied load should pass through the centroid of the connector pattern, but in many cases this is not

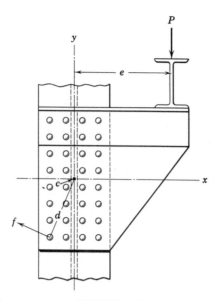

FIGURE 5-9

possible. A typical example is shown in Figure 5-9. The system tends to turn about the centroid C of the connector pattern. The purpose of this section is to determine the forces acting on the individual connectors, especially the connector carrying the largest load.

Lets begin by assuming that the connectors are numbered serially beginning at 1. Then the forces acting on the individual connectors will be F_1, F_2, F_3, ... These forces act in the direction perpendicular to a line running from the centroid C to the center of the connector on which it acts. Also let d_1, d_2, d_3, ... be the distances from the centroid to the connectors, whereas e is the eccentricity which is the perpendicular distance between the centroid and the line of action of the applied load P.

Since the joint is in static equilibrium the sum of the moments about the centroid must be zero.

$$M_C = 0$$

Therefore,

$$Pe = F_1d_1 + F_2d_2 + F_3d_3 \ldots$$

Furthermore, it can be proven that the forces due to eccentricity on the connectors are proportional to their distances from the centroid.

$$F \propto d$$

$$F = kd$$

where k is the constant of proportionality.

109

Substituting kd in the moment equation for F gives us

$$Pe = kd_1^2 + kd_2^2 + kd_3^2 \ldots$$

solving for k

$$k = \frac{Pe}{d_1^2 + d_2^2 + d_3^2 \ldots} = \frac{Pe}{\Sigma d^2}$$

It is easier to work with the x and y components of the forces F so we can restate the value of k as

$$k = \frac{Pe}{\Sigma x^2 + \Sigma y^2}$$

and since $F = kd$, the x component of F will be

$$F_x = \frac{Pey}{\Sigma x^2 + \Sigma y^2} \qquad (5\text{-}7)$$

and the y component of F will be

$$F_y = \frac{Pex}{\Sigma x^2 + \Sigma y^2} \qquad (5\text{-}8)$$

In addition to the force just determined, which is the result of the turning effect of the applied load, the connectors also have the direct force of the applied load acting on them. This they divide among themselves equally. It is

$$F_{\text{direct}} = \frac{P}{N} \qquad (5\text{-}9)$$

It is economically good practice to have all rivets or bolts in a joint the same size even though they may carry considerably different loads. In most cases it is possible to determine which fastener of a joint will carry the greatest load by visual examination. It would be the fastener furthest from the centroid of

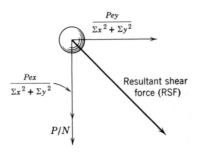

FIGURE 5-10

the fastener pattern on which the turning forces and direct forces in either direction add to each other. For example, the top right fastener in Figure 5-9 would carry the greatest load, as can be seen in Figure 5-10, which shows the forces acting on it.

The total shear force acting on a fastener is the *resultant shear force* found by

$$RSF = \sqrt{(\Sigma F_x)^2 + (\Sigma F_y)^2}$$

Example

customary

Determine the greatest force acting on a rivet in the joint in Figure 5-11 when e is 12 in., x is 3 in. y is 5 in., and P is 8000 lb.

metric

Determine the greatest force acting on a rivet in the joint in Figure 5-11 when $e = 300\,mm$, $x = 75\,mm$, $y = 125\,mm$, *and P is 8000 kg.*

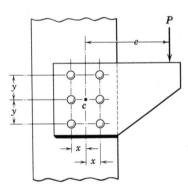

FIGURE 5-11

Solution

customary

$$\Sigma x^2 = 6\,(3^2) = 54 \text{ in.}^2$$
$$\Sigma y^2 = 4\,(5^2) = 100 \text{ in.}^2$$

The rivets carrying the greatest loads are the corner rivets. For any one of them

metric

$$\Sigma x^2 = 6\,(0.075\,m)^2 = 3.38 \times 10^{-2}\,m^2$$
$$\Sigma y^2 = 4\,(0.125\,m)^2 = 6.25 \times 10^{-2}\,m^2$$

The rivets carrying the greatest loads are the corner rivets. For any one of

111

$y = 5$ and $x = 3$. We find F_x by Equation 5-7.

$$F_x = \frac{Pey}{\Sigma x^2 + \Sigma y^2} = \frac{8000 \text{ lb } (12 \text{ in.})(5 \text{ in.})}{54 \text{ in.}^2 + 100 \text{ in.}^2}$$
$$= 3120 \text{ lb}$$

By Equation 5-8 we find

$$F_y = \frac{Pex}{\Sigma x^2 + \Sigma y^2} = \frac{8000 \text{ lb } (12 \text{ in.})(3 \text{ in.})}{54 \text{ in.}^2 + 100 \text{ in.}^2}$$
$$= 1870 \text{ lb}$$

The direct force on the rivet is

$$F_{direct} = \frac{P}{N} = \frac{8000 \text{ lb}}{6 \text{ rivets}} = 1330 \text{ lb/rivet}$$

The total force acting on a rivet in the y direction is

$$F_{y(total)} = 1870 \text{ lb} + 1330 \text{ lb} = 3200 \text{ lb}$$

The maximum shearing force, MSF, would act on one of the corner rivets. Referring to Figure 5-10 we can see that this force would be

$$MSF = \sqrt{(3200 \text{ lb})^2 + (3120 \text{ lb})^2} = 4470 \text{ lb}$$

them $y = 125$ mm and $x = 75$ mm. We find F_x by Equation 5-7.

$$F_x = \frac{Pey}{\Sigma x^2 + \Sigma y^2}$$
$$= \frac{8000 \text{ kg } (9.81 \text{ N/kg})(0.300 \text{ m})(0.125 \text{ m})}{3.38 \times 10^{-2} \text{ m}^2 + 6.25 \times 10^{-2} \text{ m}^2}$$
$$F_x = 30\,600 \text{ N}$$

By Equation 5-8 we find

$$F_y = \frac{Pex}{\Sigma x^2 + \Sigma y^2}$$
$$= \frac{8000 \text{ kg } (9.81 \text{ N/kg})(0.300 \text{ m})(0.075 \text{ m})}{3.38 \times 10^{-2} \text{ m}^2 + 6.25 \times 10^{-2} \text{ m}^2}$$
$$F_y = 18\,300 \text{ N}$$

The direct force on the rivet is

$$F_{direct} = \frac{P}{N} = \frac{8000 \text{ kg } (9.81 \text{ N/kg})}{6 \text{ rivets}}$$
$$= 13\,100 \text{ N/rivet}$$

The total force acting on a rivet in the y direction is

$$F_{y(total)} = 18\,300 \text{ N} + 13\,100 \text{ N} = 31\,400 \text{ N}$$

The maximum shearing force, MSF, would act on one of the corner rivets. Referring to Figure 5-10 we can see that the

$$MSF = \sqrt{(30\,600 \text{ N})^2 + (31\,400 \text{ N})^2}$$
$$= 43\,800 \text{ N}$$

5-5 WELDED JOINTS

Welding is any process in which two pieces of metal are fused together by heating. The most common process used in structural work is manual shielded metal-arc welding. In this process an electric arc is produced between the end of a coated metal electrode and the pieces to be welded. The arc heats the metal causing it to melt along with the metal of the welding rod, which acts as the electrode. The pool of molten metal formed by this process solidifies to form the weld which, if the proper rod and technique is used, will be as strong as the metal pieces being joined. While the melting is in process, the coating on the welding rod forms a gaseous shield which prevents impurities in the atmosphere from entering the weld.

We will consider three kinds of welds: plug, penetration, and fillet. *Plug welds* are made by drilling a hole or forming a slot in one piece of metal being joined, lapping that piece over the other, and then making the joint by filling or partially filling the hole or slot with weld metal, the weld metal fusing with both pieces being joined. Plug welds will withstand loads in shear only, with that load being equal to the shear strength of the weld for the cross-sectional area of the hole or slot. *Penetration welds*, a few types of which are shown in Figure 5-12, are formed by notching in one manner or another the parts to be joined (usually end to end) and filling the notch with weld metal as in the figure. A properly made *complete penetration weld* in which the weld passes completely through the joint is as strong as the unwelded parts being joined, making the designing of the weld unnecessary. *Fillet welds* are welds which are formed in the angle between the parts being joined as in Figure 5-13. In the succeeding sections we will study two types of loading of fillet welds—*concentrically loaded fillet welds*, where the line of action of the load passes through the centroid of the weld pattern, and *eccentrically loaded fillet welds*, where the line of action of the load does not pass through the centroid of the weld pattern.

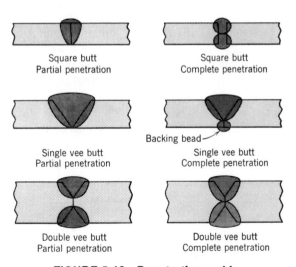

Square butt
Partial penetration

Square butt
Complete penetration

Single vee butt
Partial penetration

Backing bead

Single vee butt
Complete penetration

Double vee butt
Partial penetration

Double vee butt
Complete penetration

FIGURE 5-12 Penetration welds.

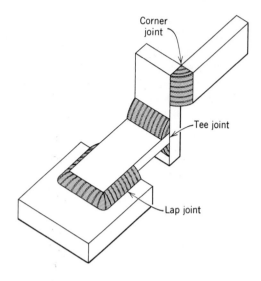

FIGURE 5-13   Fillet welds.

5-6   CONCENTRICALLY LOADED FILLET WELDS

Fillet welds may be classified as longitudinal or transverse as shown in Figure 5-14. It is generally recommended that they not be used together unless it is necessary to attain the required joint strength. Since the weld metal is elastic it can logically be assumed that a welded joint will ultimately fail in shear.

The *throat* of a fillet weld is the shortest distance from the root of the weld (where the surfaces adjacent to the weld come together) to the surface of the weld as shown in Figure 5-15. Since the sides of the weld are usually equal, the throat is 0.707t where t is the weld size as shown in the figure. The *throat area*, equal to the throat times the length of the weld, is the area that is used

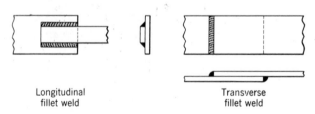

Longitudinal
fillet weld

Transverse
fillet weld

FIGURE 5-14

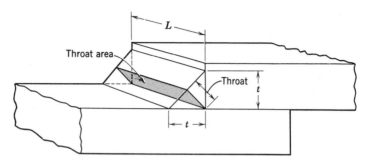

FIGURE 5-15

as the shear area. We can determine the stress in the weld by

$$f_v = \frac{P}{A} = \frac{P}{0.707tL} \qquad (5\text{-}10)$$

where P is the load on the weld, t is the size of the weld, and L is its length.

Example

customary

Determine the stress developed in a 3/8 in. weld, 5 in. long by a load of 2000 lb.

metric

Determine the stress developed in a 10 mm weld, 15 cm long by a force of 10 000 N.

Solution

customary

$$f_v = \frac{P}{0.707tL} = \frac{2000 \text{ lb}}{0.707\,(0.375 \text{ in.})(5 \text{ in.})}$$
$$= 1510 \text{ psi}$$

metric

$$f_v = \frac{P}{0.707tL} = \frac{10\,000 \text{ N}}{0.707\,(0.01 \text{ m})(0.15 \text{ m})}$$
$$= 9.43 \text{ MPa}$$

The more common problem is to determine the length of weld required to withstand a given load. Equation 5-10 can be restated for this purpose as

$$L = \frac{P}{0.707tF_v} \qquad (5\text{-}11)$$

where F_v is the allowable shear stress for the particular welding rod used. Equation 5-11 has been used to develop Tables 5-1C and 5-1M, which give the strength of fillet welds per inch in pounds and per centimeter in Newtons, respectively.

Table 5-1C

Strength per Inch of Fillet Weld—lb

Size of Weld	Electrode Strength Level—ksi			
	60	70	80	100
in.	$F_v = 18$ ksi	$F_v = 21$ ksi	$F_v = 24$ ksi	$F_v = 30$ ksi
1/8	1590	1860	2120	2650
3/16	2390	2780	3180	3980
1/4	3180	3710	4240	5300
5/16	3980	4640	5300	6630
3/8	4780	5570	6370	7960
1/2	6360	7420	8480	10600

Table 5-1M

Strength per Millimeter of Fillet Weld—N

Size of Weld	Electrode Strength Level—MPa			
	400	500	600	700
mm	$F_v = 120$ MPa	$F_v = 150$ MPa	$F_v = 180$ MPa	$F_v = 210$ MPA
3	255	318	382	445
4	339	424	509	594
5	424	530	636	742
8	679	848	1020	1190
10	848	1060	1270	1480
15	1270	1590	1910	2230

The use of these tables will simplify the determination of the required length of the welds. This is demonstrated by the following examples.

Example

customary

Determine the length of longitudinal fillet weld required for the joint in Figure 5-16. *P* is 10,000 lb, t_1 is 1/4 in., t_2 is 3/8 in., and the welding rod has a strength level of 60 ksi.

metric

Determine the length of longitudinal fillet weld required for the joint in Figure 5-16. P is 50 000 N, t_1 is 5 mm, t_2 is 10 mm, and the welding rod has a strength level of 400 MPa.

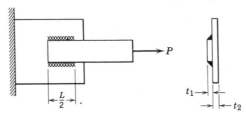

FIGURE 5-16

Solution

customary

It is customary to use a weld of the same size as the thickness of the thinnest part being joined up to about 3/8 in. The thinner plate in our example is 1/4 in. From Table 5-1C we see that a 1/4 in. weld has a shear strength of 3180 lb/in. for a welding rod of 60 strength level.

$$L = \frac{10,000 \text{ lb}}{3180 \text{ lb/in.}} = 3.14 \text{ in.}$$

Use 1-5/8 in. on each side of the plate as shown in the figure.

metric

It is common to use a weld of the same size as the thickness of the thinnest part being joined up to about 10 mm. The thinner plate in our example is 5 mm. From Table 5-1M we see that a 5 mm weld with a 400 MPa welding rod has a strength of 424 N/mm. Thus

$$L = \frac{50\,000 \text{ N}}{424 \text{ N/mm}} = 118 \text{ mm.}$$

Use 6 cm on each side of the plate as shown in the figure.

5-7   ECCENTRICALLY LOADED FILLET WELDS

A general treatment of eccentrically loaded fillet joints is beyond the scope of this book. The discussion here will be limited to cases in which the line of action of the force lies in the same plane as the welds, or cases where the line

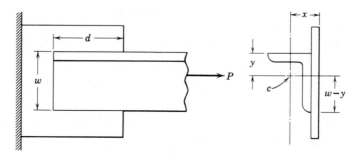

FIGURE 5-17

of action of the force lies in a plane parallel to the plane of the welds at a distance too small to cause any significant stress due to that distance. Such a case is shown in Figure 5-17 where we have an angle which is to be welded to the plate in the position shown. The load P acts through the centroid of the angle.

The first step in analyzing this joint is to draw a free-body diagram of the angle as in Figure 5-18, which shows the force P in equilibrium with the resisting forces F_1 and F_2 of the welds at either side of the joint. Since this is a system in equilibrium we can now determine the force F_1 by summing moments about a point on the line of action of force F_2.

$$M_{F_2} = F_2(0) - F_1 w + P(w - y) = 0$$

Solving for F_1 we find

$$F_1 = \frac{P(w - y)}{w}$$

The length of the weld of resisting force F_1 is

$$L_1 = \frac{F_1}{\text{strength/unit length of weld}}$$

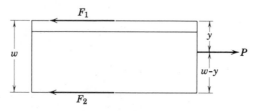

FIGURE 5-18

118

Substituting the above value of F_1 in this last equation gives us

$$L_1 = \left(\frac{P}{\text{strength/unit length of weld}}\right)\left(\frac{w - y}{w}\right)$$

The first term in the above equation,

$$\frac{P}{\text{strength/unit length of weld}},$$

is actually the total length of weld required. Letting it be L we have

$$L_1 = L\frac{w - y}{w} \qquad\qquad (5\text{-}12)$$

In the same way we find the length of the weld needed to provide the resistance F_2 by summing moments about F_1 to be

$$L_2 = L\frac{y}{w} \qquad\qquad (5\text{-}13)$$

Example

customary

Determine the sizes of the longitudinal fillet welds required for the joint in Figure 5-17 if the angle is an L6 × 4 × 1/2, the welds are 3/8 in. with a strength level of 80, and P is 40,000 lb.

metric

Determine the sizes of the longitudinal fillet welds required for the joint in Figure 5-17 if the angle is a 150 × 90 × 10. The welds are 8 mm in size with a strength level of 500 and P is 180 kN.

Solution

customary

From Table 5-1C we find the strength/in. of weld to be 6370 lb.

$$L = \frac{P}{\text{strength/in. of weld}} = \frac{40{,}000\text{ lb}}{6370\text{ lb/in.}}$$
$$= 6.28\text{ in.}$$

Because the long side of the angle lies on the plate, $w = 6$ in. From Appendix 2 we

metric

From Table 5-1M we find the strength/cm of weld to be 8480 N.

$$L = \frac{P}{\text{strength/mm of weld}} = \frac{180\,000\text{ N}}{848\text{ N/mm}}$$
$$= 212\text{ mm}$$

Since the long side of the angle lies on the plate, w = 150 mm. From Appendix

find y to be 1.99 in. Substituting these values in Equation 5-12 gives us

$$L_1 = L \frac{w - y}{w} = 6.28 \text{ in.} \frac{6 - 1.99}{6} = 4.20 \text{ in.}$$

Substituting the same values in Equation 5-13 gives us

$$L_2 = L \frac{y}{w} = 6.28 \text{ in.} \frac{1.99}{6}$$
$$= 2.08 \text{ in.}$$

2 we find y to be 50 mm. Substituting these values in Equation 5-12 gives us

$$L_1 = L \frac{w - y}{w} = 212 \text{ mm} \frac{150 \text{ mm} - 50 \text{ mm}}{150 \text{ mm}}$$
$$= 141 \text{ mm}$$

Substituting these same values in Equation 5-13 gives us

$$L_2 = L \frac{y}{w} = 212 \text{ mm} \frac{50 \text{ mm}}{150 \text{ mm}}$$
$$= 70.7 \text{ mm, use 71 mm}$$

Occasionally there is not enough space to provide sufficient longitudinal fillet weld to satisfy the strength requirements of a welded joint. In such cases it is necessary to use a transverse weld as shown in Figure 5-19. When a

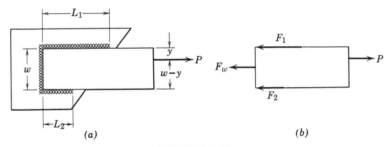

(a)   (b)

FIGURE 5-19

transverse weld is used it will usually be layed across the entire end. Its resisting force F_w will act from its center as shown in the free-body diagram of the bar in Figure 5-19b. The determination of the required lengths L_1 and L_2 is made by summing moments as in the previous derivation which gives us, for a joint with a full transverse weld, using the same terminology as before,

$$L_1 = L \frac{w - y}{w} - \frac{w}{2} \qquad (5\text{-}14)$$

and

$$L_2 = L \frac{y}{w} - \frac{w}{2} \qquad (5\text{-}15)$$

Example

customary

Determine the lengths L_1 and L_2 of the joint in Figure 5-19 when P is 40,000 lb, w is 4 in., y is 1 in., and the weld is 1/4 in. with a strength level of 60 ksi.

metric

Determine the lengths L_1 and L_2 of the joint in Figure 5-19 when P is 180 kN, w is 100 mm, y is 25 mm, and the weld is 5 mm with a strength level of 400.

Solution

customary

$$L = \frac{P}{\text{strength/in.}} = \frac{40{,}000 \text{ lb}}{3180 \text{ lb/in.}} = 12.6 \text{ in.}$$

Using Equation 5-14

$$L_1 = L\frac{w-y}{2} - \frac{w}{2} = 12.6 \text{ in.}$$

$$\times \frac{4 \text{ in.} - 1 \text{ in.}}{4 \text{ in.}} - \frac{4 \text{ in.}}{2} = 7.45 \text{ in.}$$

Using Equation 5-15

$$L_2 = L\frac{y}{w} - \frac{w}{2} = 12.6 \text{ in.} \frac{1 \text{ in.}}{4 \text{ in.}} - \frac{4 \text{ in.}}{2}$$

$$= 1.15 \text{ in.}$$

In practice you would specify L_1 to be 7-1/2 in. and L_2 to be 1-1/4 in.

metric

$$L = \frac{P}{\text{strength/in.}} = \frac{180 \text{ kN}}{424 \text{ N/mm}}$$
$$= 425 \text{ mm}$$

Using Equation 5-14

$$L_1 = L\frac{w-y}{w} - \frac{w}{2} = 425 \text{ mm}$$

$$\times \frac{100 \text{ mm} - 25 \text{ mm}}{100 \text{ mm}} - \frac{100 \text{ mm}}{2}$$

$$L_1 = 269 \text{ mm}$$

Using Equation 5-14

$$L_2 = L\frac{y}{w} - \frac{w}{2} = 425 \text{ mm}$$

$$\times \frac{25 \text{ mm}}{100 \text{ mm}} - \frac{100 \text{ mm}}{2} = 56 \text{ mm}$$

In practice you would specify L_1 to be 270 mm and L_2 to be 60 mm.

5-8 ACCOMPLISHMENT CHECKLIST

From your studies of this chapter do you now know

1. **The difference between a lap joint and a butt joint, a main plate and a cover plate, a continuous joint and a structural joint, a framed connection and a seated connection.**

2. The three principal ways in which a bolted or riveted joint is likely to fail if overloaded.

3. How to determine the shearing, bearing, and tensile stresses in bolted or riveted lap and butt joints.

4. How to determine the maximum shearing stresses in eccentrically loaded bolted and riveted joints.

5. The meaning of the terms *welding* and *manual shielded metal-arc welding*.

6. The purpose of the coating on a welding rod.

7. The difference between plug, penetration, and fillet welds.

8. How to determine the stresses in concentrically loaded fillet welds.

9. How to determine the size and amount of fillet weld required to withstand a given load.

10. How to determine the size and location of fillet welds for eccentrically loaded joints where the line of action of the load lies in or near the same plane as the welds?

5-9 SUMMARY

In a *lap joint* the pieces to be joined are lapped over one another and bolted or riveted or welded in that position. In a *butt joint* the plates to be joined are butted together end to end. If welded, they are usually notched and then welded directly. If bolted or riveted, plates called *cover plates* are placed over the joint extending on each side with the bolts or rivets fastening the joint by penetrating both the cover and main plates. A *continuous joint* is one in which the rivet pattern is continuously repeated over a considerable distance. A *structural joint* is one used to connect structural members such as channels, beams, and columns. In a *framed connection* a member is suspended between a pair of angles, whereas in a *seated connection* the supported member sits on an angle which is attached to the supporting member.

In riveted or bolted joints *shear failure* occurs when the fasteners are sheared off, *bearing failure* occurs when the material behind the fasteners crushes, and *tensile failure* occurs when the metal tears across a section of the joint.

The shearing stress in a connector is determined by the relationship

$$f_s = \frac{P}{NA} \qquad (5\text{-}1)$$

where f_s is the shearing stress, P is the total load on the joint, N is the number of connectors, and A is the cross-sectional area of each connector.

The bearing stress in the material behind the connectors is determined by

$$f_c = \frac{P}{NtD} \qquad (5\text{-}2)$$

where f_c is the bearing (compressive) stress in the material behind the connectors, P is the total load on the joint, N is the number of connectors, t is the thickness of the member being analyzed, and D is the diameter of the connectors.

The tensile stress in the section of a main or cover plate is determined by

$$f_t = \frac{P}{(b - N_s D_h)t} \qquad (5\text{-}3)$$

where f_t is the tensile stress, b is the width of the section, N_s is the number of bolt or rivet holes in the section, t is the thickness of the section, and D_h is the diameter of the holes.

In a butt joint the connectors are in *double shear*, the load tending to shear them simultaneously at two sections. The stress in double shear is

$$f_{vb} = \frac{P}{2NA} \qquad (5\text{-}4)$$

The main plate bearing stress in a butt joint is

$$f_{cm} = \frac{P}{NDt} \qquad (5\text{-}5)$$

The cover plate bearing stress in a butt joint is

$$f_{cc} = \frac{P}{2NDt} \qquad (5\text{-}6)$$

When the line of action of an imposed load does not pass through the centroid of the fasteners of a joint, the fastener furthest from the centroid will experience the greatest force and hence the greatest stress. The force on this fastener may be determined by separately finding the force due to the direct pull of the load and that caused by the turning effect of the load caused by its eccentricity. This turning effect is best handled if it is further broken up into its x and y components. The equations for the turning components are

$$F_x = \frac{Pey}{\Sigma x^2 + \Sigma y^2} \qquad (5\text{-}7)$$

$$F_y = \frac{Pex}{\Sigma x^2 + \Sigma y^2} \qquad (5\text{-}8)$$

where F_x is the x component due to turning effect, F_y is the y component, P is the total load, e is the eccentricity of the load, y is the y component of the distance of the line of action from the centroid of the fasteners, x is the x component, Σx^2 is the sum of the squares of the x distances of the fasteners from the centroid, and Σy^2 is the sum of the squares of the y distances. The direct force is given by

$$F_{\text{direct}} = \frac{P}{N} \qquad (5\text{-}9)$$

Welding is any process in which two pieces of metal are fused together by heating. A common process is *manual shielded metal-arc welding* in which an electric arc is produced between the end of a coated rod (electrode) and the pieces to be welded. The coating on the welding rod is vaporized by the heat and forms a gaseous shield, preventing impurities in the atmosphere from entering the weld. *Plug welds* are made by drilling a hole or forming a slot in one piece of metal being joined, lapping that piece over the other piece, and filling or partially filling the hole or slot with weld metal. *Penetration welds* are formed by notching the ends of the parts being joined and filling the notch with weld metal. *Fillet welds* are formed in the angle between two adjoining members as in Figure 5-13, page 114.

The *throat* of a fillet weld is the shortest distance from the root of the weld to the surface of the weld. The *throat area* is the throat times the length of the weld. Fillet welds are assumed to fail in shear. The value of that shear is

$$f_v = \frac{P}{A} = \frac{P}{0.707tL} \qquad (5\text{-}10)$$

where f_v is the shearing stress, P the total load, t the thickness of the plate, and L the length of the weld. When the length of weld is to be determined the relationship is

$$L = \frac{P}{0.707tF_v} \qquad (5\text{-}11)$$

When the welds cannot be placed symmetrically with respect to the line of action of the load, the required length of longitudinal welds sufficient to withstand the load can be determined by

$$L_1 = L \frac{w - y}{w} \tag{5-12}$$

and

$$L_2 = L \frac{y}{w} \tag{5-13}$$

where L_1 and L_2 are the lengths of the welds, w is the width of the member being welded, y is the distance from the edge of the member to the line of action of the force—all distances being as shown in Figure 5-17, page 118.

When a full transverse weld is used in conjunction with longitudinal welds, the lengths of the longitudinal welds are determined by

$$L_1 = L \frac{w - y}{w} - \frac{w}{2} \tag{5-14}$$

and

$$L_2 = L \frac{y}{w} - \frac{w}{2} \tag{5-15}$$

PROBLEMS

lap joint problems

5-1C Determine the shear, bearing, and tensile stresses in the lap joint shown when P is 8000 lb, b is 2 in., t is 1/4 in., and the diameter of the bolt is 3/4 in. Holes are punched.

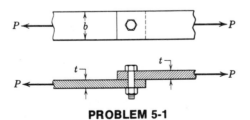

PROBLEM 5-1

5-1M *Determine the shear, bearing, and tensile stresses in the lap joint shown when P is 30 000 N, b is 50 mm, t is 5 mm, and the diameter of the bolt is 20 mm. Holes are punched.*

5-2C   Determine the load at which the joint would fail if the straps being joined are 3003-H 16 aluminum with a maximum tensile stress F_{tu} of 26 ksi and a maximum bearing stress F_{cu} of 49 ksi, and the rivets are 2024-T4 with an ultimate shear stress F_{vu} of 38 ksi. The rivets are 1/8 in. diameter, b is 1 in., and t is 1/16 in. Recommended hole diameter is 0.1285 in.

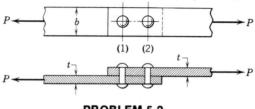

PROBLEM 5-2

5-2M   *Determine the load at which the joint would fail if the straps being joined are 3003-H 16 aluminum with a maximum tensile stress F_{tu} of 180 MPa and a maximum bearing stress F_{cu} of 330 MPa, and the rivets are 2024-T4 aluminum with an ultimate shear stress F_{vu} of 260 MPa. The rivets are 3 mm diameter, b is 25 mm, and t is 1.5 mm. Recommended hole diameter is 3.085 mm.*

5-3C   Determine the tensile, bearing, and shear stresses in the joint shown when P is 15 kips, b is 3 in., t is 3/8 in., and the bolts are 7/8 diameter. Holes are drilled.

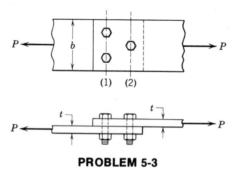

PROBLEM 5-3

5-3M   *Determine the tensile, bearing, and shear stresses in the joint shown when P is 60 kN, b is 75 mm, t is 10 mm, and the bolts are 20 mm. Holes are drilled.*

5-4C   Determine the shear, bearing, and tensile stresses in the continuous aluminum lap joint shown when p is 0.75 in., t is 0.05 in., the rivets are 0.125 diameter, and P is 600 lb/ft. *Hint:* Examine a section of the joint one pitch p wide.

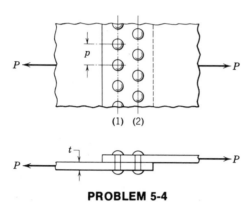

PROBLEM 5-4

5-4M   *Determine the shear, bearing, and tensile stresses in the continuous aluminum lap joint shown when p is 20 mm, t is 1.20 mm, the rivets are 3 mm diameter, and P is 9000 N/m. Hint: Examine a section of the joint one pitch p wide.*

butt joint problems

5-5C   Determine the shear, bearing, and tensile stress in the joint shown if b is 2 in., t_m is 3/8 in., t_c is 1/4 in., P is 20 kip, and the bolts are 3/4 in. Holes are punched.

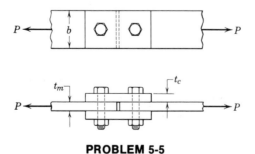

PROBLEM 5-5

5-5M   *Determine the shear, bearing, and tensile stress in the joint shown if b is 50 mm, t_m is 10 mm, t_c is 5 mm, P is 90 kN, and the bolts are 20 mm. Holes are punched.*

5-6C   Determine the maximum allowable load P on the joint shown if both the main and cover plates are A36 steel with F_t 22 ksi, b is 3-1/2 in., t_m is 1/2 in., t_c is 1/4 in., the bolts are 3/4 in. diameter, and P is 60 kips. The bolts are A490 with an allowable shear stress of 20 ksi. Holes are drilled. How could the design of the joint be improved?

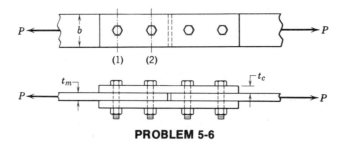

PROBLEM 5-6

5-6M   *Determine the maximum allowable load P on the joint shown if both the main and cover plates are A36 steel with F_t 150 MPa, b is 90 mm, t_m is 10 mm, t_c is 5 mm, the bolts are 20 mm diameter, and P is 80 kN. The bolts are A490 with an allowable shear stress of 140 MPa. Holes are drilled. How could the design of the joint be improved?*

5-7C   Determine the shear, bearing, and tensile stresses in the joint shown if P is 50,000 lb, t_m is 1/2 in., t_c is 1/4 in., b is 4 in., and the diameter of the bolts is 5/8 in. Holes are punched.

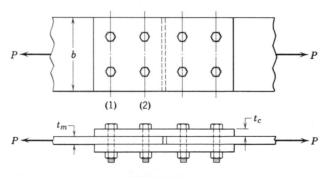

PROBLEM 5-7

5-7M   Determine the shear, bearing, and tensile stresses in the joint shown if P is 200 kN, t_m is 15 mm, t_c is 7.5 mm, b is 100 mm, and the diameter of the bolts is 20 mm. Holes are punched.

5-8C   Determine the shear, bearing, and tensile stresses in the joint shown if P is 60,000 lb, t_m is 1/2 in., t_c is 5/16 in., b is 5 in., and the diameter of the rivets is 7/8 in. Holes are punched.

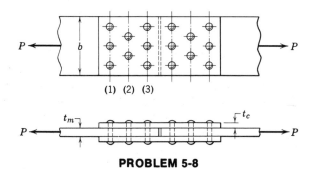

PROBLEM 5-8

5-8M   Determine the shear, bearing, and tensile stresses in the joint shown if P is 250 kN, t_m is 15 mm, t_c is 10 mm, b is 125 mm, and the diameter of the rivets is 20 mm. Holes are punched.

5-9C   Determine the allowable load for the joint shown if all plates are A572 steel with F_t of 30 ksi, and the bolts are 3/4 in. A490 with F_v of 20 ksi, b is 6 in., t_m is 5/8 in., and t_c is 3/8 in. The holes are punched.

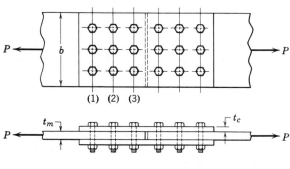

PROBLEM 5-9

5-9M *Determine the allowable load for the joint shown if all plates are A572 steel with F_t of 200 MPa, the bolts are 20 mm A490 with F_v of 140 MPa, b is 150 mm, t_m is 15 mm, and t_c is 10 mm. The holes are punched.*

5-10C Determine the allowable load for the joint shown if all plates are A572 steel with F_t of 30 ksi, the rivets are 1 in. A490 with F_v of 20 ksi, b is 8 in., t_m is 1 in. and t_c is 3/8 in. The holes are punched.

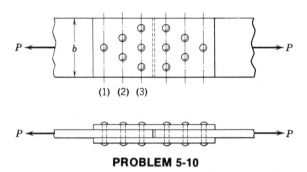

PROBLEM 5-10

5-10M *Determine the allowable load for the joint shown if all plates are A572 steel with F_t of 200 MPa, the rivets are 25 mm A490 with F_v of 140 MPa, b is 200 mm, t_m is 25 mm, and t_c is 10 mm. The holes are punched.*

eccentrically loaded joint problems

5-11C Determine the shearing stress in the bolts in the figure if P is 100 lb, b is 6 in., and a is 2 in.

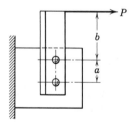

PROBLEM 5-11

5-11M *Determine the shearing stress in the bolts in the figure if P is 500 N, b is 150 mm, and a is 50 mm.*

5-12C   Determine the size of bolts required for the joint shown if P is 1000 lb, b is 24 in., and a is 8 in. The bolts are all to be the same size and are to be A490 with an allowable shear stress of 18,000 psi.

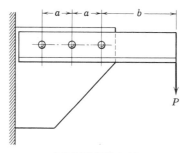

PROBLEM 5-12

5-12M   *Determine the size of bolts required for the joint shown if P is 1000 kg, b is 1.2 m, and a is 0.2 m. The bolts are all to be the same size and are to be A490 with an allowable shear stress of 125 MPa.*

5-13C   Determine the maximum stress developed in the bolts of the joint shown if the load P is 20 kips and the bolts are 1-1/4 in. in diameter. The distance b is 3 ft and a is 4 in.

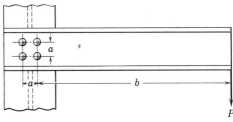

PROBLEM 5-13

5-13M   *Determine the maximum stress developed in the bolts of the joint shown if the load P is 4000 kg and the bolts are 32 mm in diameter. The distance b is 1 m and a is 120 mm.*

5-14C   The cantilever beam of Problem 5-13 supports a load P of 1 ton. Determine the size of the rivets required to support this load if they are A502 Grade 1 hot-driven with an F_v of 15 ksi, a is 4 in. and b is 6 ft.

131

5-14M *The cantilever beam of Problem 5-13 supports a load P of 1 ton. Determine the size of the rivets required to support this load if they are A502 Grade 1 hot-driven with an F_v of 100 MPa, a is 100 mm, and b is 2 m.*

5-15C Determine the maximum shearing stress in the bolted joint in the figure if P is 2000 lb, a is 5 in., b is 20 in., and the bolts are 1/2 in. diameter.

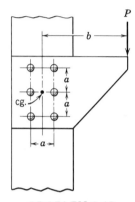

PROBLEM 5-15

5-15M *Determine the maximum shearing stress in the bolted joint in the figure if P is 10 kN, a is 125 mm, b is 500 mm, and the diameter of the bolts is 15 mm.*

5-16C Determine the maximum shearing force on the bolts of the joint in the figure if P is 40,000 lb, a is 6 in., b is 24 in., and θ is 60°.

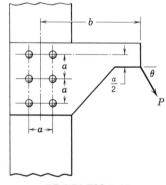

PROBLEM 5-16

5-16M   *Determine the maximum shearing force on the bolts of the joint in the figure if P is 180 kN, a is 150 mm, b is 600 mm, and θ is 60°.*

5-17C   Determine the maximum distance *b* for the joint shown if *P* is 8000 lb, *a* is 5 in., and the bolts are 3/4 A490 with an allowable shear stress of 18 ksi.

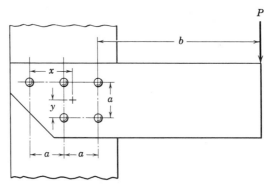

PROBLEM 5-17

5-17M   *Determine the maximum distance b for the joint shown if P is 35 kN, a is 125 mm, and the bolts are 20 mm diameter A490 with an allowable shear stress of 124 MPa.*

5-18C   Determine the maximum shear stress in the bolts in the joint shown if *P* is 6000 lb, *a* is 2-1/2 in., and the bolts are 1/2 in. diameter.

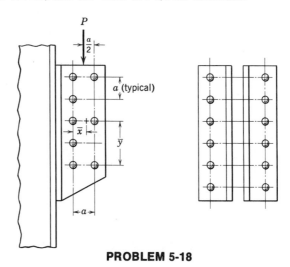

PROBLEM 5-18

CONNECTIONS

5-18M *Determine the maximum shear stress in the bolts in the joint shown if P is 25 kN, a is 60 mm, and the bolts are 12 mm in diameter.*

concentrically loaded fillet weld problems

5-19C Determine the shearing stress in a concentrically loaded fillet weld if the load is 15,000 lb, the weld is 1/4 in., and its length is 6 in.

5-19M *Determine the shearing stress in a concentrically loaded fillet weld if the load is 50 kN, the weld is 5 mm, and its length is 150 mm.*

5-20C Determine the length of 3/8 in. fillet weld required for a concentrically loaded joint which supports a 50,000 lb load if F_v for the weld is 24 ksi.

5-20M *Determine the length of 10 mm fillet weld required for a concentrically loaded joint which supports a 200 kN load if F_v for the weld is 180 MPa.*

5-21C The load on a concentrically loaded fillet welded joint is 20,000 lb. The maximum length available is 6 in. Determine the smallest size of weld of electrode strength level 70 that can be used.

5-21M *The load on a concentrically loaded fillet welded joint is 90 kN. The maximum length available is 150 mm. Determine the smallest size of weld of electrode strength level 500 that can be used.*

5-22C Determine the maximum load P which can be supported by the system in the figure if the fillet welds are 1/4 in. with electrode strength level 60 and a is 6 in. No transverse weld is to be used.

5-22M *Determine the maximum load P which can be supported by the system in the figure if the fillet welds are 5 mm with electrode strength level 400 and a is 150 mm. No transverse weld is to be used.*

5-23C If the system in Problem 5-22C was designed to support a load of 240,000 lb and the fillet welds were 5/16 in. thick and 10 in. long, what strength level welding rod would be required?

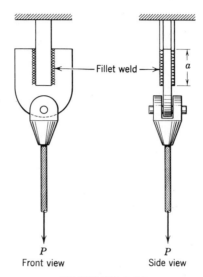

PROBLEM 5-22

5-23M *If the system in Problem 5-22M was designed to support a load of 980 kN and the fillet welds were 8 mm thick and 250 mm long, what strength level welding rod would be required?*

eccentrically loaded fillet weld problems

5-24C An L3 × 3 × 3/8 angle carries a 40,000 lb load acting through its centroid. It is to be attached to its support by 5/16 in. longitudinal fillet welds using electrode strength level 70 welding rod. Determine the size and location of the required welds.

5-24M *A 70 × 70 × 8 angle carries a 180 kN load acting through its centroid. It is to be attached to its support by 8 mm longitudinal fillet welds using electrode strength level 500 welding rod. Determine the size and location of the required welds.*

5-25C An L5 × 3 × 3/8 angle carries a 60,000 lb load acting through its centroid. It is to be attached with its long side down by 5/16 in. longitudinal fillet welds using a welding rod of strength level 80. Determine the size and location of the required welds.

5-25M   *A 125 × 75 × 8 angle carries a 270 kN load acting through its centroid. It is to be attached with its long side down by 8 mm longitudinal welds using a welding rod of strength level 600. Determine the size and location of the required welds.*

5-26C   Same as 5-24C except that both a full transverse weld and longitudinal welds are used.

5-26M   *Same as 5-24M except that both a full transverse weld and longitudinal welds are used.*

5-27C   Same as 5-25C except that both a full transverse weld and longitudinal welds are used.

5-27M   *Same as 5-25M except that both a full transverse weld and longitudinal welds are used.*

5-28C   Plate B is to be welded using 1/4 in. thick longitudinal fillet welds of rod strength level 60 to plate A. Determine the length of the required welds if a is 6 in., b is 3 in, c is 5 in., and P is 20 kips.

5-28M   *Plate B is to be welded using 5 mm thick longitudinal fillet welds of rod strength level 400 to plate A. Determine the length of the required welds if a is 150 mm, b is 75 mm, c is 125 mm, and P is 90 kN.*

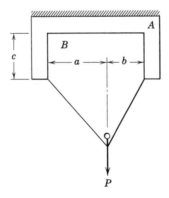

PROBLEM 5-28

5-29C   Same as Problem 5-28C except a 3 in. transverse weld is to be used on the right side (dimension *b*) of the plate B in addition to the longitudinal fillet welds.

5-29M   *Same as Problem 5-28M except a 75-mm transverse weld is to be used on the right (dimension b) of plate B in addition to the longitudinal fillet welds.*

5-30C   A load *P* is attached to the vertex of the triangular plate A. Plate A is welded to Plate B by $\frac{1}{2}$ in. fillet welds on the sides of the triangular plate. Determine the length and location of the welds if a rod of strength level 80 is used and *P* is 75,000 lb, *a* is 6 in., *b* is 3 in., *c* is 9 in., and *d* is 6 in.

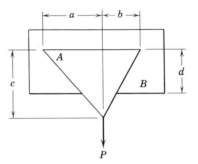

PROBLEM 5-30

5-30M   *A load P is attached to the vertex of the triangular plate A. Plate A is welded to Plate B by 15 mm fillet welds on the sides of the triangular plate. Determine the length and location of the welds if a rod of strength level 600 is used, P is 350 kN, a is 150 mm, b is 75 mm, c is 225 mm, and d is 150 mm.*

CHAPTER
OBJECTIVES

The purpose of studying this chapter is to

1. Learn how to determine the stresses in solid and hollow circular shafts.

2. Learn how to determine the stresses developed in shafts which are transmitting power between members of machines.

3. Learn how to determine the stresses in the connectors of shaft couplings and how to determine the number and size of connectors needed to transmit given amounts of power.

4. Learn how to determine the amount of strain energy in a given shaft when it is twisted.

5. Learn how torsion bars can act as springs and how to determine their effect as springs.

CHAPTER ——————————— 6
TORSION

In this chapter we will consider the effects of pure torsion. *Pure torsion* occurs when a machine or structural member is simply twisted with no tensile, compressive or bending forces involved.

6-1 TORSION OF CIRCULAR SHAFTS

When a torque is applied to a shaft as illustrated in Figure 6-1 it twists the shaft about its longitudinal axis. When the torque is pure torque the centerline will remain perfectly straight and will not change in length. If that is true, then the only forces resisting the applied torque act parallel to the plane of the applied torque and perpendicular to the axis of the shaft. We see then that the only stresses set up by the rotational deformation of the shaft are shearing stresses as shown in Figure 6-1. Also as the center of the shaft is approached the

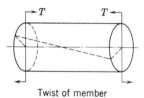
Twist of member

Rotation of
cross section

Direction of
shear stresses

Distribution of
shear stresses

FIGURE 6-1

deformation becomes less until at the center it is zero. The infinitesimal (immeasurably small) stress on any area of the shaft is directly proportional to its distance from the axis. This proportionality can be stated algebraically as

$$\frac{f_v}{R} = \frac{f_{va}}{y}$$

where f_v is the maximum shear stress (at the surface of the shaft), R is the radius of the shaft, and f_{va} is the stress on any infinitesimally small area at a distance y from the axis of the shaft.

From the above equation the stress on any immeasurably small area of the cross section of a shaft with a torque applied to it is

$$f_{va} = \frac{f_v y}{R}$$

The force acting on this infinitesimal area dA is $f_{va}dA$ or using the value of f_{va} from the above equation it is

$$\frac{f_v y}{R} dA$$

This force provides a part of the resisting torque in the shaft.

$$T_a = F_a y = \frac{f_v y}{R} dAy = \frac{f_v}{R} y^2 dA$$

where T_a is the resisting torque of an infinitesimal area dA, F_a is the resisting force of that area, y is the distance from the axis of the shaft to that area, f_v is the maximum shearing stress in the shaft, and R is the radius of the shaft.

The total resisting torque, which must be the same as the applied torque, is the sum of the resisting torques of all the infinitesimal areas that make up the cross section of the shaft, or

$$T = \sum T_a = \sum \frac{f_v}{R} y^2 dA$$

Since f_v and R are constants the above equation can be written

$$T = \frac{f_v}{R} \sum y^2 dA$$

The quantity $\sum y^2 dA$ is the polar moment of inertia J of the cross section about its center. Therefore

$$T = \frac{f_v J}{R} \qquad (6\text{-}1)$$

The above equation can be used to determine the torque for a shaft of given radius and shearing stress.

For a solid shaft the polar moment of inertia is

$$J = \frac{\pi R^4}{2}$$

For a hollow circular shaft the polar moment of inertia is

$$J = \frac{\pi}{2}(R^4 - r^4)$$

where r is the inside radius of the hollow shaft and R is the outside radius.

Sometimes it is difficult to find information concerning the shear strength and yield strength in shear of steel. The ultimate shear strength ranges from 2/3 to 3/4 of the tensile strength. The yield stress in shear is approximately equal to its theoretical value of $1/\sqrt{3}$ times the yield stress in tension.*

The following examples will demonstrate the application of Equation 6-1.

Example

customary

Determine the maximum allowable torque for a 1-15/16 diameter shaft made of SAE 1045 steel with an allowable shear stress of 55,000 psi.

metric

Determine the maximum allowable torque for a 50 mm diameter shaft made of SAE 1045 steel with an allowable shear stress of 380 MPa.

Solution

customary

$$R = \frac{1\text{-}15/16 \text{ in.}}{2} = \frac{1.937 \text{ in.}}{2} = 0.969 \text{ in.}$$

$$J = \frac{\pi R^4}{2} = \frac{\pi (0.969 \text{ in.})^4}{2} = 1.385 \text{ in.}^4$$

$$T = \frac{f_v J}{R} = \frac{55,000 \text{ psi} (1.385 \text{ in.}^4)}{0.969 \text{ in.}}$$
$$= 78,600 \text{ lb-in.}$$

metric

$$R = \frac{50 \text{ mm}}{2} = 25 \text{ mm} = 0.025 \text{ m}$$

$$J = \frac{\pi R^4}{2} = \frac{\pi (0.025 \text{ m})^4}{2} = 6.14 \times 10^{-7} \text{ m}^4$$

$$T = \frac{f_v J}{R} = \frac{380 \text{ MPa} (6.14 \times 10^{-7} \text{ m}^4)}{0.025 \text{ m}}$$
$$= 9\,330 \text{ N} \cdot m$$

*Based on Energy-of-Distortion (Hencky-von Mises) Yield Criterion. See Fred B. Seely and James O. Smith, *Advanced Mechanics of Materials*, 2nd ed. (New York: John Wiley & Sons, Inc., 1952), pp. 76–91.

Example

customary

Determine the maximum shearing stress in a shaft having an outside diameter of 1.5 in. and an inside diameter of 1 in. when a torque of 25,000 lb-in. is applied to it.

metric

Determine the maximum shearing stress in a shaft having an outside diameter of 40 mm and an inside diameter of 10 mm when a torque of 3000 N·m is applied to it.

Solution

customary

$R = 1.5 \text{ in.}/2 = 0.75 \text{ in.}$,

$r = 1 \text{ in.}/2 = 0.50 \text{ in.}$

$J = \dfrac{\pi}{2}(R^4 - r^4) = \dfrac{\pi}{2}((0.75 \text{ in.})^4 - (0.50 \text{ in.})^4)$

$= 0.399 \text{ in.}^4$

From Equation 5-1 we find

$f_v = \dfrac{TR}{J} = \dfrac{25{,}000 \text{ lb-in. } (0.75 \text{ in.})}{0.399 \text{ in.}^4}$

$= 47{,}000 \text{ psi}$

metric

$R = 20 \text{ mm} = 0.020 \text{ m}$,

$r = 5 \text{ mm} = 0.005 \text{ m}$

$J = \dfrac{\pi}{2}(R^4 - r^4) = \dfrac{\pi}{2}((0.020 \text{ m})^4$

$- (0.005 \text{ m})^4) = 2.5 \times 10^{-7} \text{ m}^4$

From Equation 5-1 we find

$f_v = \dfrac{TR}{J} = \dfrac{3000 \text{ N·m } (0.020 \text{ m})}{2.5 \times 10^{-7} \text{ m}^4}$

$= 240 \text{ MPa}$

There are times when the diameter of the shaft is to be determined. In such cases the allowable stress F_v is used in place of f_v and the value of J is substituted in Equation 6-1 as follows.

$$T = \frac{F_v J}{R} = \frac{F_v(\pi R^4/2)}{R} = \frac{F_v \pi R^3}{2} = \frac{F_v \pi (D/2)^3}{2}$$

Solving for R

$$R = \sqrt[3]{\frac{2T}{\pi F_v}} \quad \text{or} \quad D = \sqrt[3]{\frac{16T}{\pi F_v}} \tag{6-2}$$

Example

customary

Determine the diameter of a shaft made of SAE 4340 steel with an allowable

metric

Determine the diameter of a shaft made of SAE 4340 steel with an

shearing stress of 150,000 psi to handle a torque of 300,000 lb-in.

allowable shearing stress of 1 000 MPa to handle a torque of 35 000 N·m.

Solution

customary

$$D = \sqrt[3]{\frac{16T}{\pi F_v}} = \sqrt[3]{\frac{16\,(300{,}000\ \text{lb-in.})}{\pi\,(150{,}000\ \text{psi})}}$$

$$= \sqrt[3]{10.19\ \text{in.}^3} = 2.17\ \text{in.}$$

Use a 2-3/16 in. diameter shaft.

metric

$$D = \sqrt[3]{\frac{16T}{\pi F_v}} = \sqrt[3]{\frac{16\,(35\,000\ N{\cdot}m)}{\pi\,(1\,000\ MPa)}}$$

$$= 0.056\,3\ m$$

Use a 60 mm shaft.

6-2 POWER TRANSMISSION

The principal use of shafting is to transmit power from one member of a machine to another, from one part of a machine to another, or from one machine to another. Power is the rate at which work is performed. The mechanical unit of power in the customary system is the horsepower, one horsepower being 550 ft-lb/sec. In the metric system the unit of power is the watt, one watt being one J/s. Because watt is a very small unit, the kilowatt is a more practical unit to use.

In order to determine the torsional shearing stresses in rotating shafts it is necessary to know the relationship between power and torque. In Figure 6-2 we see the cross section of a shaft being turned by a force F. For each

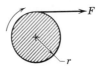

FIGURE 6-2

revolution of the shaft the work is

$$U = Fd = F(2\pi R) \quad \text{or} \quad U = 2\pi FR$$

The power (work per second) is

$$\text{Power} = 2\pi FR(N)$$

where N is the number of revolutions per second and FR is the torque; therefore the power equation may be written

$$\text{Power} = 2\pi NT$$

One horsepower (hp) in the customary system of units equals 396,000 in.-lb/min, the horsepower is

$$\text{hp} = \frac{2\pi NT}{396,000 \text{ in.-lb/min}} \qquad \text{Customary System} \qquad (6\text{-}3\text{C})$$

where N is the number of revolutions per minute and T is the torque in in.-lb.

In the metric system the unit of power, the watt, is one newton-meter per second so the power is

$$W = \omega T \qquad \text{Metric System} \qquad (6\text{-}3\text{M})$$

where ω is the number of radians per second and T is the torque in N·m.

To determine the stress induced in a shaft during the transmission of power we need only solve either of the preceding equations for T and use that value in Equation 6-1. The equations for torque are

$$T = \frac{(396,000 \text{ in.-lb/min})(\text{hp})}{2\pi N} \qquad \text{for customary units} \qquad (6\text{-}4\text{C})$$

and

$$T = \frac{W}{\omega} \qquad \text{for metric units} \qquad (6\text{-}4\text{M})$$

Example

customary	metric
Determine the maximum shearing stress in a 2 in. shaft when transmitting 30 hp at 200 rpm.	*Determine the stress in a 50 mm shaft when transmitting 25 kW at 200 rpm.*

Solution

customary

$$T = \frac{(396,000 \text{ in.-lb/min})(\text{hp})}{2\pi N}$$

$$= \frac{396,000 \text{ in.-lb/min}\,(30 \text{ hp})}{2\pi\,200 \text{ rpm}}$$

$$T = 9450 \text{ lb-in.}$$

metric

$$\omega = \frac{2\pi\,(200 \text{ rpm})}{60 \text{ sec/min}} = 20.9 \text{ rad/s}$$

$$T = \frac{W}{\omega} = \frac{25 \text{ kW}}{(20.9 \text{ rad/s})} = 1\,200 \text{ N·m}$$

Substitute this in Equation 5-1 after finding J

$$J = \frac{\pi R^4}{2} = \frac{\pi (1 \text{ in.})^4}{2} = 1.57 \text{ in.}^4$$

$$f_v = \frac{TR}{J} = \frac{9450 \text{ lb-in. } (1 \text{ in.})}{1.57 \text{ in.}^4}$$

$$= 6020 \text{ psi}$$

Substitute this in Equation 5-1 after finding J

$$R = 25 \text{ mm} = 0.025 \text{ m}$$

$$J = \frac{\pi R^4}{2} = \frac{\pi (0.025 \text{ m})^4}{2} = 6.14 \times 10^{-7} \text{ m}^4$$

$$f_v = \frac{TR}{J} = \frac{1\,200 \text{ N} \cdot \text{m } (0.025 \text{ m})}{6.14 \times 10^{-7} \text{ m}^4}$$

$$= 48.9 \text{ MPa}$$

6-3 SHAFT COUPLINGS

Power may be taken off of a shaft in a number of ways; the most common being by the use of pulleys, gears, or couplings. The coupling is used where the power is to be transmitted to a unit which revolves concentrically with the shaft. This could be another shaft or some other power chain member such as the wheel of an automobile.

We will concern ourselves here with the most simple coupling, the flanged coupling, such as is illustrated in Figure 6-3. In this case the flange is turned on the end of the shaft as is frequently done with small precision parts. Larger flanges are usually made separately from the shaft and are then welded or attached to the shaft by some other method. The most likely modes of failure of flanged couplings are shearing where the flange joins the shaft, shearing of the bolts, crushing of the bolts, or crushing of the metal of the flange behind the bolts. Good design would provide that the bolts fail before the flange itself fails because the bolts would be less costly to replace in case of catastrophic failure.

We will first consider the shearing of the flange at the surface of the shaft. The area sheared would be

$$A = 2\pi Rt$$

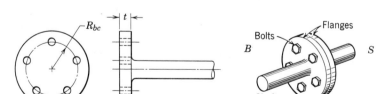

FIGURE 6-3

where R is the radius of the shaft and t is the flange thickness. The force acting on this surface is

$$P = \frac{T}{R}$$

where T is the torque being transmitted by the shaft.

The shearing stress in the flange at the surface of the shaft would be

$$F_{vf} = \frac{P}{A} = \frac{T/R}{2\pi Rt} = \frac{T}{2\pi R^2 t} \qquad (6\text{-}5)$$

To determine the shearing stress on the bolts we must first determine the force acting parallel to the surface of the shaft at the bolt circle radius.

$$P_{bc} = \frac{T}{R_{bc}}$$

where T is the torque being transmitted by the shaft and R_{bc} is the radius of the bolt circle of the flange.

Each bolt will take an equal share of the total force or P_{bc}/n where n is the number of bolts. The shearing stress in each bolt is P_{bc}/n divided by the cross-sectional area of a bolt. Thus the shearing stress on the bolts is

$$f_{vb} = \frac{P_{bc}/n}{A_{bolt}} = \frac{(T/R_{bc})/n}{\pi D_b^2/4} = \frac{4T}{\pi R_{bc} n D_b^2} \qquad (6\text{-}6)$$

where T is the torque being transmitted by the shaft, R_{bc} is the radius of the bolt circle, n is the number of bolts, and D_b is the diameter of the bolts.

The average bearing stress at the surface of a bolt is the same as that of the area of the flange behind it with which it is in contact. It is equal to the load on the bolt divided by the cross-sectional area of the hole perpendicular to the direction of the load.

$$f_c = \frac{P_{bc}}{t D_b}$$

or in terms of the total area supporting the load, the bearing stress on the bolts and the flange surfaces behind the bolts is

$$f_c = \frac{P_{bc}}{nt D_b} = \frac{T/R_{bc}}{nt D_b} = \frac{T}{nt D_b R_{bc}} \qquad (6\text{-}7)$$

Example

customary

The shaft in Figure 6-3 transmits 40 hp at 250 rpm. Determine the number of bolts required and the minimum flange thickness. The specifications are as follows:

Shaft diameter D	2 in.
Bolt circle radius R_{bc}	2 in.
Bolt diameter D_b	3/8 in.
Allowable shear stress for flange F_{vf}	15,000 psi
Allowable shear stress for bolt F_{vb}	18,000 psi
Allowable bearing stress for flange F_{cf}	20,000 psi
Allowable bearing stress for bolt F_{cb}	24,000 psi

metric

The shaft in Figure 6-3 transmits 30 kW at 250 rpm. Determine the number of bolts required and the minimum flange thickness. The specifications are as follows.

Shaft diameter D	*50 mm*
Bolt circle radius R_{bc}	*50 mm*
Bolt diameter D_b	*10 mm*
Allowable shear stress for flange F_{vf}	*100 MPa*
Allowable shear stress for bolt F_{vb}	*125 MPa*
Allowable bearing stress for flange F_{cf}	*140 MPa*
Allowable bearing stress for bolts F_{cb}	*165 MPa*

Solution

customary

Using Equation 6-4C we find

$$T = \frac{(396{,}000 \text{ in.-lb/min})(hp)}{2\pi N}$$
$$= \frac{(396{,}000 \text{ in.-lb/min})(40 \text{ hp})}{2\pi (250 \text{ rpm})}$$
$$= 10{,}100 \text{ lb-in.}$$

Now determine the number of bolts required by solving for n, using Equation 6-6. The number of bolts

$$n = \frac{4T}{\pi R_{bc}D_b^2 F_{v_b}} = \frac{4(10{,}100 \text{ lb-in.})}{\pi (2 \text{ in.})(0.375 \text{ in.})^2(18{,}000 \text{ psi})}$$
$$= 2.54 \text{ bolts}$$

Use three bolts.

metric

Using Equation 6-4M we find

$$T = \frac{W}{\omega} = \frac{W}{2\pi N} = \frac{30 \text{ kW}}{2\pi (250 \text{ rpm})/60 \text{ s/min}}$$
$$= 1150 \text{ N}\cdot m$$

Now determine the number of bolts required by solving for n using Equation 6-6. The number of bolts

$$n = \frac{4T}{\pi R_{bc}D_b^2 F_{v_b}}$$
$$= \frac{4(1150 \text{ N}\cdot m)}{\pi (0.050 \text{ m})(0.010 \text{ m})^2(125 \text{ MPa})}$$
$$= 2.34 \text{ bolts}$$

Use three bolts.

Having determined the number of bolts, we can now determine the required thickness of the flange. Solving Equation 6-7 for the flange thickness, t gives us

$$t = \frac{T}{nF_{c_f}D_bR_{bc}}$$

$$= \frac{10{,}100 \text{ lb-in.}}{3 \text{ bolts } (20{,}000 \text{ psi})(0.375 \text{ in.})(2 \text{ in.})}$$

$$= 0.224 \text{ in.}$$

We must now determine if this flange thickness is sufficient to withstand shearing of the flange at the shaft surface. We can do this by substituting the above value in Equation 6-5.

$$f_{v_f} = \frac{T}{2\pi R^2 t} = \frac{10{,}100 \text{ lb-in.}}{2\pi (1 \text{ in.})^2 (0.224 \text{ in.})}$$

$$= 7180 \text{ psi}$$

Since this stress is less than the allowable 15,000 psi, the flange thickness of 0.224 in. is satisfactory.

Having determined the number of bolts, we can now determine the required thickness of the flange. Solving Equation 6-7 for the flange thickness t, gives us

$$t = \frac{T}{nF_{c_f}D_bR_{bc}}$$

$$= \frac{1150 \text{ N·m}}{3 \text{ bolts } (140 \text{ MPa})(0.010 \text{ m})(0.050 \text{ m})}$$

$$= 5.48 \text{ mm}$$

We must now determine if this flange thickness is sufficient to withstand shearing of the flange at the shaft surface. We can do this by substituting the above value in Equation 6-5.

$$f_{v_f} = \frac{T}{2\pi R^2 t} = \frac{1150 \text{ N·m}}{2\pi (0.025)^2 (0.005\,48 \text{ m})}$$

$$= 53.4 \text{ MPa}$$

Since this stress is less than the allowable, (100 MPa), the flange thickness of 5.48 mm is satisfactory.

6-4 ANGLE OF TWIST OF ROUND SHAFTS

Frequently it is necessary to determine the extent of the twisting of shafts under loads. For example the deflection of shafts by twisting can have an effect on the operation of gears because it can change the relationship of the mating gear teeth. Torsion bar springs are basically bars that absorb energy in the same manner as ordinary springs but are dependent on the elastic twisting of the bars to perform their function.

To develop the equation needed to determine the angle of twist of round shafts we will examine the piece of shaft in Figure 6-4. Here we have a piece of round shaft rigidly fixed at one end with a torque applied at the other end. Before twisting, the solid line ab was drawn on the surface. Upon application of the torque the section of shaft twists so that point b advances to b'. The angle of twist is θ. The distance from b to b' is $R\theta$, where R is the radius of the

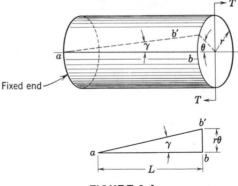

FIGURE 6-4

shaft and θ is the angle of twist in radians. The diagram below the illustration of the shaft shows the unit deformation of the shaft as the angle γ which in radian measure is $R\theta/L$, where L is the length of the shaft. Also, as with lineal deformations, the angular deformation is proportional to the stress. The constant of proportionality in this case is called the *shear modulus of elasticity G*. It is called a shear modulus because the stress developed by twisting is a shearing stress.

The angular deformation is

$$\gamma = \frac{f_v}{G}$$

Equating the two values of γ gives us

$$\frac{R\theta}{L} = \frac{f_v}{G} \quad \text{or} \quad \theta = \frac{f_v L}{RG} \tag{6-8}$$

From Equation 6-1

$$f_v = \frac{TR}{J}$$

so

$$\theta = \frac{(TR/J)L}{RG} = \frac{TL}{JG} \tag{6-9}$$

where T is the applied torque, L is the length of the shaft, J is the polar moment of inertia of the shaft, and G is the shear modulus of elasticity of the material of the shaft.

Example

customary	**metric**

customary

Determine the angle of twist of a 1-15/16 in. shaft 6 ft long when transmitting 20 hp at 600 rpm. The modulus of elasticity in shear is 11,400,000 psi.

metric

Determine the angle of twist of a 50 mm shaft 2 m long when transmitting 15 kW at 60 rad/s. The modulus of elasticity in shear is 78.6 GPa.

Solution

customary

$$T = \frac{(396,000 \text{ in.-lb/min})(hp)}{2\pi N}$$

$$= \frac{(396,000 \text{ in.-lb/min})(20 \text{ hp})}{2\pi \, 600 \text{ rpm}}$$

$$T = 2100 \text{ in.-lb}$$

$$J = \frac{\pi R^4}{2} = \frac{\pi (1\text{-}15/16 \div 2 \text{ in.})^4}{2} = 1.38 \text{ in.}^4$$

$$\theta = \frac{TL}{JG} = \frac{2100 \text{ in.-lb} (72 \text{ in.})}{1.38 \text{ in.}^4 (11,400,000 \text{ psi})}$$

$$= 0.00961 \text{ rad}$$

or

$$\theta = 0.00961 \text{ rad} \left(\frac{360}{2\pi} \text{ degree/radian} \right)$$

$$= 0.551°$$

metric

$$T = \frac{W}{\omega} = \frac{15 \text{ kW}}{60 \text{ rad/s}} = 250 \text{ N} \cdot \text{m}$$

$$J = \frac{\pi R^4}{2} = \frac{\pi (.025 \text{ m})^4}{2} = 6.14 \times 10^{-7} \text{ m}^4$$

$$\theta = \frac{TL}{JG} = \frac{250 \text{ N} \cdot \text{m} (2 \text{ m})}{6.14 \times 10^{-7} \text{ m}^4 (78.6 \text{ GPa})}$$

$$= 0.0104 \text{ rad}$$

or

$$\theta = 0.0104 \text{ rad} \left(\frac{360}{2\pi} \text{ degree/radian} \right)$$

$$= 0.596°$$

6-5 TORSION BARS

As mentioned in the previous section a round shaft can be used as an energy absorption device as in the torsion bar suspension system of an automobile.

Example

customary

The space available limits the length of the torsion bar in Figure 6-5 to 10.5 in. A force F of 10,000 lb is applied to the lever at a distance *d* of 12 in. Determine the required minimum diameter of the torsion bar if it is made of SAE 1045 steel with an allowable shearing stress of 90,000 psi and a *G* of 11,400,000 psi. How far will the lever move if the torsion bar is made of shafting to the nearest 1/16-in. above the determined minimum size? The shaft is supported to permit angular deflection only.

metric

The space available limits the length of the torsion bar in Figure 6-5 to 270 mm. A force F of 45,000 N is applied to the lever at a distance d of 0.300 m. Determine the required minimum diameter of the torsion bar if it is made of SAE 1045 steel with an allowable shearing stress of 620 MPa and a G of 78.6 GPa. How far will the lever move if the torsion bar is made of shafting to the nearest even mm above the determined minimum size? The shaft is supported to permit angular deflection only.

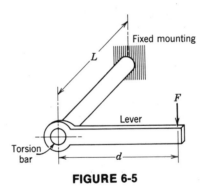

FIGURE 6-5

Solution

customary

$T = Fd = 10,000$ lb $(12$ in.$) = 120,000$ lb-in.
Use Equation 6-2 to determine the shaft diameter

$$D = \sqrt[3]{\frac{16T}{f_v}} = \sqrt[3]{\frac{16\,(120,000\text{ lb-in.})}{90,000\text{ psi}}}$$

$$= 2.77 \text{ in.}$$

Use a $2\frac{13}{16}$ shaft for the torsion bar.

metric

$T = Fd = 45\,000$ N $(0.300$ m$)$
   $= 13\,500$ N·m

Use Equation 6-2 to determine the shaft diameter

$$D = \sqrt[3]{\frac{16T}{f_v}} = \sqrt[3]{\frac{16\,(13\,500\text{ N·m})}{620\text{ MPa}}}$$

$$= 0.070\,4 \text{ m}$$

Use a 72 mm shaft for the torsion bar.

The angle of twist can now be easily found by use of Equation 6-8.

$$\theta = \frac{f_v L}{RG}$$

Our shaft diameter would be 2.81 in. making $R = 1.41$ in.

$$\theta = \frac{90{,}000 \text{ psi} (10.5 \text{ in.})}{1.41 \text{ in.} (11{,}400{,}000 \text{ psi})} = 0.0588 \text{ rad}$$
$$= 3.37°$$

The angle of twist can now be easily found by use of Equation 6-8.

$$\theta = \frac{f_v L}{RG} = \frac{620 \text{ MPa} (0.270 \text{ m})}{0.036 \text{ m} (78.6 \text{ GPa})}$$
$$= 0.059\,2 \text{ rad} = 3.39°$$

6-6 ACCOMPLISHMENT CHECKLIST

From your studies of this chapter do you now understand

1. How to determine the maximum allowable torque for any given round shaft whether solid or hollow.

2. How to determine the required diameter of shaft to handle a given torque or power.

3. How to determine the shearing stress in shafts transmitting power when in pure torsion.

4. How to determine the stresses in members of shaft couplings.

5. How to determine the required dimensions for a flanged coupling to transmit a given amount of power.

6. The relationship between torsional stress and strain and the modulus of elasticity in shear.

7. The principle of the torsion bar.

8. How to determine the proper size of torsion bar to absorb a given amount of energy when the bar is mounted so that it is in pure torsion?

6-7 SUMMARY

Pure torsion occurs when a machine or structural member is simply twisted with no tensile, compressive, or bending forces exerted on it.

The torque in a shaft is related to the shearing stress by the equation

$$T = \frac{f_v J}{R} \qquad (6\text{-}1)$$

where f_v is the shearing stress, J is the polar moment of inertia of the shaft, and R is the radius of the shaft.

The relationship for determining the required radius of a shaft to transmit a given torque with a shaft having a certain allowable shearing stress is

$$R = \sqrt[3]{\frac{2T}{\pi F_v}} \quad \text{or} \quad D = \sqrt[3]{\frac{16T}{\pi F_v}} \qquad (6\text{-}2)$$

The relationships of horsepower and torque in the customary system are

$$\text{hp} = \frac{2\pi NT}{396,000 \text{ in.-lb/min}} \quad \text{or} \quad T = \frac{396,000 \text{ (in.-lb/min)(hp)}}{2\pi N} \qquad (6\text{-}3C \text{ and } 4C)$$

where hp is the horsepower, T the torque in in.-lb, and N is rpm. The relationships of horsepower and torque in the customary system are

$$W = \omega T \quad \text{or} \quad T = \frac{W}{\omega} \qquad (6\text{-}3M \text{ and } 4M)$$

where W is watts, ω is radians per second, and T is torque in newton-meters.

The shearing force in the flange of a flanged coupling at the outside diameter of the shaft is

$$f_{v_f} = \frac{T}{2\pi R^2 t} \qquad (6\text{-}5)$$

where T is the torque, R is the radius of the shaft, and t is the thickness of the flange.

The shearing stress in the bolts of a flanged coupling is

$$f_{v_b} = \frac{4T}{\pi R_{bc} n D_b^2} \qquad (6\text{-}6)$$

where T is the shaft torque, R_{bc} is the radius of the bolt circle, n is the number of bolts, and D_b is the diameter of the bolts.

153

The compressive stress on the bolts and on the flange surfaces behind the bolts is

$$f_c = \frac{T}{ntD_bR_{bc}} \tag{6-7}$$

For a flanged coupling all three of the stresses obtained by the preceding equations must be within the allowable limits.

In the same manner that compressive and tensile stresses are proportional to the strain so are the shearing stresses proportional to the angular deformation. The constant of proportionality for shear is called the *shear modulus of elasticity* and is represented by the symbol G.

The deformation of a shaft by pure torsion is called the angle of twist θ, whereas the deformation per unit of length of the shaft is given the symbol γ. The total deformation of a shaft of length L, stress f_v, radius r, and shear modulus of elasticity G is

$$\theta = \frac{f_v L}{RG} \tag{6-8}$$

By substituting TR/J for f_v in the above equation a very useful equation is obtained as follows

$$\theta = \frac{TL}{JG} \tag{6-9}$$

A bar being twisted within its elastic range absorbs energy. In this manner it is acting like a spring and is called a *torsion bar*. The energy load is usually transmitted to the torsion bar by a lever as shown in Figure 6-5.

PROBLEMS

torsion in solid circular shafts problems

6-1C   Determine the maximum allowable torque for a 1-3/16 diameter shaft made of SAE 3140 steel with an allowable shear stress of 60,000 psi.

6-1M   *Determine the maximum allowable torque for a 30 mm diameter shaft made of SAE 3140 steel with an allowable shear stress of 415 MPa.*

6-2C   The copper shaft in an explosion-proof piece of equipment has an allowable shear strength of 15,000 psi and must transmit 40,000 lb-in. of torque. Determine the required diameter of the shaft.

6-2M *The copper shaft in an explosion-proof piece of equipment has an allowable shear strength of 100 MPa and must transmit 4500 N·m of torque. Determine the required diameter of the shaft.*

6-3C Determine the shearing stress in a 2-in. diameter shaft when it is transmitting 100,000 lb-in. of torque.

6-3M *Determine the shearing stress in a 50 mm diameter shaft when it is transmitting 10 000 N·m of torque.*

torsion in hollow circular shafts problems

6-4C Determine the maximum allowable torque for a hollow shaft 2.00 in. in O.D. with a 0.250 in. wall made of 3003-H14 aluminum having an allowable shear strength of 9000 psi.

6-4M *Determine the maximum allowable torque for a hollow shaft 50 mm in O.D. with a wall thickness of 5 mm if the shaft is made of 3003-H14 aluminum having an allowable shear strength of 60 MPa.*

6-5C A hollow steel shaft is required to transmit 50,000 lb-in. of torque. Determine the inside and outside diameters of the shaft if the I.D. is three quarters of the O.D. and the shaft is made of SAE 3140 steel having an allowable shear strength of 60,000 psi.

6-5M *A hollow steel shaft is required to transmit 6000 N·m of torque. Determine the inside and outside diameters of the shaft if the I.D. is three quarters of the O.D. and the shaft is made of SAE 3140 steel having an allowable shear strength of 415 MPa.*

6-6C A hollow steel drive shaft transmits 10,000 lb-in. torque. It is made of SAE 1045 steel with an allowable shearing strength of 55,000 psi. Determine the minimum wall thickness required if the O.D. is 3.00 in.

6-6M *A hollow steel drive shaft transmits 1000 N·m torque. It is made of SAE 1045 steel with an allowable shearing strength of 380 MPa. Determine the minimum wall thickness if the O.D. is 70 mm.*

6-7C   Determine the diameter of a solid steel shaft that will transmit the same torque as a hollow steel shaft of 2-3/16 in. O.D. and 1/4-in. wall thickness of the same material. Which weighs the most per foot and what is the ratio of the weight of the solid shaft to that of the hollow shaft?

6-7M   *Determine the diameter of a solid steel shaft that will transmit the same torque as a hollow steel shaft of 80 mm O.D. and 10 mm wall thickness of the same material. Which weighs the most and what is the ratio of the weight of the solid shaft to that of the hollow shaft?*

power transmission problems

6-8C   Determine the maximum shearing stress in a 1 in. shaft transmitting 15 hp at 750 rpm.

6-8M   *Determine the maximum shearing stress in a 30 mm shaft transmitting 10 kW at 10 rad/s.*

6-9C   Determine the maximum horsepower that can safely be transmitted by a 1-7/16 in. diameter SAE 1045 steel shaft when rotating at 3000 rpm if the allowable stress of the steel is 55,000 psi.

6-9M   *Determine the maximum watts that can safely be transmitted by a 36 mm diameter SAE 1045 steel shaft having an allowable stress of 380 MPa when rotating at 300 rad/s.*

6-10C   Determine the minimum diameter for a shaft made of SAE 4340 steel having an allowable shearing stress of 90,000 psi when the shaft is transmitting 250 hp at 1250 rpm.

6-10M   *Determine the minimum diameter for a shaft made of SAE 4340 steel having an allowable shearing stress of 600 MPa when the shaft is transmitting 200 kW at 20 rad/s.*

6-11C   Determine the maximum shearing stress in a hollow steel shaft having an O.D. of 0.375 in. and a wall thickness of 0.010 in. when transmitting 0.50 hp at 4000 rpm.

6-11M   *Determine the maximum shearing stress in a hollow steel shaft having an O.D. of 10 mm and a wall thickness of 0.25 mm when transmitting 375 W at 425 rad/s.*

6-12C   Determine the maximum horsepower that can safely be transmitted at 1200 rpm by an SAE 1045 steel shaft with an O.D. of 1-7/16 in. and a wall thickness of 7/32-in. if the allowable stress of the steel is 55,000 psi.

6-12M   *Determine the maximum power that can safely be transmitted at 1200 rpm by an SAE 1045 steel shaft with an O.D. of 35 mm and a wall thickness of 5 mm if the allowable stress of the steel is 400 MPa.*

shaft coupling problems

6-13C   The shaft in the figure is 2-in. in diameter and transmits 40 hp at 350 rpm. The flange is welded to the shaft and the weld has the same strength as the shaft material which in shear is for design purposes 8000 psi. Determine the minimum flange thickness required to transmit the load assuming the weld to be the weakest part of the coupling.

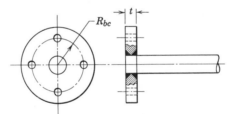

PROBLEM 6-13

6-13M   *The shaft in the figure is 50 mm in diameter and transmits 30 kW at 6 rpm. The flange is welded to the shaft and the weld has the same strength as the shaft material which in shear is for design purposes 55 MPa. Determine the minimum flange thickness required to transmit the load assuming the weld to be the weakest part of the coupling.*

6-14C   Determine the number of 3/8-in. diameter bolts on a 4 in. bolt circle required to transmit the load in the flanged coupling of Problem 6-13C. The bolts are made of SAE 1035 steel having a design stress of 7500 psi in shear and 15,000 psi in bearing.

6-14M   *Determine the number of 10 mm diameter bolts on a 100 mm bolt circle required to transmit the load in the flanged coupling of Problem 6-13M. The bolts are made of SAE 1035 steel having a design stress of 50 MPa in shear and 100 MPa in bearing.*

6-15C   A coupling is held together by three pins on a 3 in. bolt circle. It is required that they fail in single shear when the torque reaches a point no higher than 250 lb-in. so that more costly members of the machine will not be damaged by overloading. The material of the pins is to be 1100-0 aluminum with an ultimate shear strength of 9000 psi. Determine their maximum diameter.

6-15M   *A coupling is held together by three pins on a 75 mm bolt circle. It is required that they fail in single shear when the torque reaches a point no higher than 30 N·m so that more costly members of the machine will not be damaged by overloading. The material of the pins is to be 1100-0 aluminum with an ultimate shear strength of 62 MPa. Determine the required pin diameter.*

6-16C   Two shafts of 5-in. diameter are coupled by 1-1/4 in. thick plain flanges forged on their ends. The couplings are held together by six 1-1/4 diameter bolts with allowable stresses of 15,000 psi in shear and 40,000 psi in bearing on a 7-3/4 in. bolt circle. Determine the maximum torque that should be transmitted by this coupling if the shaft material is SAE 1045 steel with an allowable shear stress of 45,000 psi.

6-16M   *Two shafts of 250 mm diameter are coupled by 30 mm thick plain flanges forged on their ends. The couplings are held together by six 30 mm bolts with allowable stresses of 100 MPa in shear and 275 MPa in bearing on a 200 mm bolt circle. Determine the maximum torque that should be transmitted by this coupling if the shaft material is SAE 1045 steel with an allowable shear stress of 310 MPa in the as forged condition.*

6-17C   100 hp is to be transmitted at 1200 rpm by an SAE 4130 shaft having maximum allowable stresses in shear of 70,000 psi and in bearing 100,000 psi. This shaft is to be connected to another by flanges formed on the ends of the shafts. The flanges are to be bolted together by 6 bolts on a bolt circle having a diameter of twice the shaft diameter to the nearest 1/8 in. Determine the diameter of the shafts, the thickness of the flanges, and the size of the bolts if the bolts are SAE 1045 steel with allowable stresses in shear of 45,000 psi and in bearing of 60,000 psi.

6-17M *75 kW is to be transmitted at 125 rad/s by an SAE 4130 shaft having a maximum allowable shearing stress of 500 MPa and a maximum allowable bearing stress of 700 MPa. This shaft is to be connected to another by flanges formed on the ends of the shafts. The flanges are to be bolted together by 6 bolts on a bolt circle having a diameter of twice the shaft diameter to the nearest even millimeter. Determine the diameter of the shafts, the thickness of the flanges, and the size of the bolts if the bolts are SAE 1045 steel with allowable stresses in shear of 300 MPa and in bearing of 400 MPa.*

angle of twist problems

6-18C Determine the angle of twist of a 2 in. shaft 8 ft long when transmitting 40 hp at 350 rpm. The shear modulus of elasticity of the steel is 11,400,000 psi.

6-18M *Determine the angle of twist of a 50 mm shaft 2.50 m long when transmitting 30 kW at 40 rad/s. The shear modulus of elasticity of the steel is 78.6 GPa.*

6-19C Determine the angle of twist of a hollow steel shaft 8 ft long having an O.D. of 2 in. and an I.D. of 1 in. when transmitting 40 hp at 350 rpm. The modulus of elasticity in shear for steel is 11,400,000 psi.

6-19M *Determine the angle of twist of a hollow steel shaft 2.50 m long having an O.D. of 50 mm and an I.D. of 15 mm when transmitting 30 kW at 40 rad/s. The modulus of elasticity in shear for steel is 78.6 GPa.*

6-20C The figure shows a schematic of a gear train which is transmitting 15 hp when gears 2 and 3 are rotating at 400 rpm. Determine the largest permissible dimension x if in that length the maximum allowable twist is 0.10°. G is 11,400,000 psi. D is 1-1/2 in.

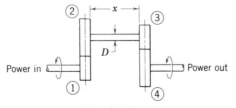

PROBLEM 6-20

TORSION

6-20M *The figure shows a schematic of a gear train which is transmitting 10 kW when gears 1 and 2 are rotating at 45 rad/s. Determine the largest permissible dimension x if in that length the maximum allowable twist is 0.002 rad. G is 78.6 GPa. D is 40 mm.*

6-21C If in Problem 6-20C the distance x was 8 in., what would be the smallest permissible diameter D?

6-21M *If in Problem 6-20M the distance x was 200 mm, what would be the smallest permissible diameter D?*

6-22C An aluminum tube 6 ft long with a 3 in. O.D. and a wall thickness of 0.188 in. is twisted to its maximum allowable shearing stress of 18,000 psi. G is 3,800,000 psi. Determine its angular deflection.

6-22M *An aluminum tube 2 m long with a 80 mm O.D. and a wall thickness of 5 mm is twisted to its maximum allowable shearing stress of 125 MPa. G is 26.2 GPa. Determine its angular deflection.*

torsion bar problems

6-23C A torsion bar 12 in. long is to be designed to absorb 80,000 lb-in. of torque per degree of twist. Determine the required diameter of the bar if it is made of steel with a modulus of elasticity in shear of 11,400,000 psi.

6-23M *A torsion bar 300 mm long is to be designed to absorb 500 kN·m per radian of twist. Determine the required diameter of the bar if it is made of steel with a G of 78.6 GPa.*

6-24C A tubular steel torsion bar has an O.D. of 1 in. and a wall thickness of 0.15 in. Determine the energy it can absorb per degree of twist if it is 20 in. long and has a G of 11,400,000 psi.

6-24M *A tubular steel torsion bar has an O.D. of 25 mm and a wall thickness of 5 mm. Determine the energy it can absorb for each mrad of twist if it is 500 mm long and has a G of 78.6 GPa.*

6-25C The figure shows a simple torque wrench in top view and cross section. The vertical tube is a hollow torsion bar. The indicator is firmly attached to the rib at the bottom of the tubular torsion bar and thus does not rotate with the tubular section. The divisions on the dial indicate tens of lb-ft of torque. Determine the angular distance between divisions and the amount of force P per division if L is 18 in., h is 8 in., D is 0.800 in., and d is 0.790 in. G is 11,400,000 psi.

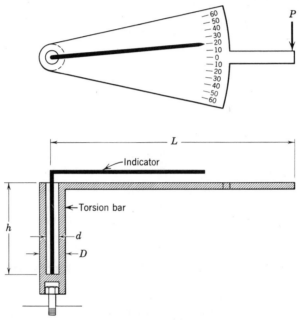

PROBLEM 6-25

6-25M The figure shows a simple torque wrench in top view and cross section. The vertical tube is a hollow torsion bar. The indicator is firmly attached to the rib at the bottom of the tubular torsion bar and thus does not rotate with the tubular section. The divisions on the dial indicate hundreds of newton-meters. Determine the angular distance between divisions and the amount of force P per division if L is 500 mm, h is 200 mm, D is 20 mm, and d is 16 mm. G is 78.6 GPa.

CHAPTER
OBJECTIVES ——————————————

To this end the object of studying this chapter is to learn

1. Some of the basic terminology relating to beams and their loading.

2. How to determine the shearing force and bending moment of a beam at any point along its axis.

3. How to draw shear and bending moment diagrams to aid in the analyzing of beams.

4. How to use the shear and bending moment diagrams to find the critical sections of beams where the maximum stresses will occur.

5. The relations between load intensity, shearing force, and bending moment.

CHAPTER————————————7
SHEAR FORCE AND
BENDING MOMENT
IN BEAMS

A *beam* is a member of a structure or machine which is designed to sustain loads acting perpendicular to its longitudinal axis. In the case of a structure the loading is usually gravitational but in a machine it is frequently applied by other bodies in directions other than that in the direction of the force of gravity. In the design of beams it is important to know the nature of the forces acting at all points in the beam.

7-1 TYPES OF BEAMS AND LOADS

Beams in which the material is uniform throughout are called *homogeneous*; those of which the flanges are made of higher strength material than the web are called *hybrid beams*, while those made of more than one material are called *composite beams*. Large beams fabricated from plates may be further designated as *girders*.

Beams may also be classified according to their support as shown in Figure 7-1. The *simple beam* is supported at the ends in such a manner that the movement of the beam due to expansion or flexure is not restricted. A *cantilever beam* is one that has one end rigidly imbedded in a support, while the other end is unsupported. An *overhanging beam* is one that is supported similarly to the simple beam except that one or both ends extend beyond the supports. The *fixed beam* has both ends rigidly supported so that the supported portions of the ends can experience no movement. The *continuous beam* extends over more than two supports. The top three beams in Figure

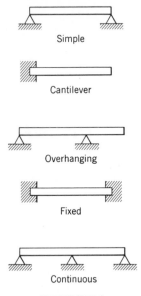

Simple

Cantilever

Overhanging

Fixed

Continuous

FIGURE 7-1

7-1 may be analyzed by the methods of statics and are therefore *statically determinate beams*. The fixed and continuous beams do not yield to analysis by statics and so are classified as *statically indeterminate beams*. In this chapter we will deal only with statically determinate beams. Statically indeterminate beams will be covered in Chapter Ten.

Loads may be classified as *stationary* or *moving*. We are here interested only in stationary loads. Stationary loads may be divided into concentrated, uniform, and variable. *Concentrated loads* are considered to act at a point, although in practice this rarely happens. In many cases the assumption that a load acting on a relatively small area acts at a point causes no significant error. When the load is distributed over a portion of the beam the load per unit of length is called the *load intensity*. A *uniform load* is one of constant load intensity. A *variable load* is one having varying intensity, usually in some definite pattern from one point on the beam to another.

7-2 SHEARING FORCE AND BENDING MOMENT

When external forces (loads) are applied to a beam they tend to shear and bend it. In shear the loads tend to cause the beam to part along a plane perpendicular to the longitudinal axis of the beam. In bending the loads on the beam tend to cause rotation of the parts of the beam about its supports. Because the beam is deformed by these loads there are stresses set up within it that determine whether or not it can safely carry the loads. The determination of these stresses is the subject of the next chapter. However, we must first be able to determine the maximum external shear forces and bending moments at any point of a beam.

In our analysis of beams we will always assume their longitudinal axes to be horizontal. The *shearing force V* on the cross section of a beam is the algebraic sum of all the applied loads and reactions to either the left or the right of the cross section. In this book we will use the convention that the shearing force is the algebraic sum of the applied loads and reactions to the *left* of the section (unless otherwise specified) with up being positive and down negative.

The *bending moment M* at any cross section perpendicular to the longitudinal axis of a beam is the sum of the moments of the loads and reactions on either side of that cross section. To avoid unnecessary complication we will use the convention that the *bending moment* is the sum of the moments of the loads and reactions to the *left* of the section with the moments of upward acting forces being positive, those downward negative.

7-3   SHEAR AND BENDING MOMENT DIAGRAMS

The shearing force (shear) and bending moment for all points of a beam can conveniently be diagrammed. The picture thus produced will reveal the points where the highest stresses will be developed and may suggest possible improvements in the loading pattern.

Example of a simple beam with a single concentrated load

Figure 7-2(a) shows a simple beam *ABC* which supports a single concentrated load. The weight of the beam itself will be assumed to be insignificant, although in most instances this is not the case. The units of the spaces and forces will be unnamed and may be considered to be either customary or metric units. Think of the weight units as being pounds or kilograms and the length units as being feet or meters. The procedure for drawing the diagrams is as follows.

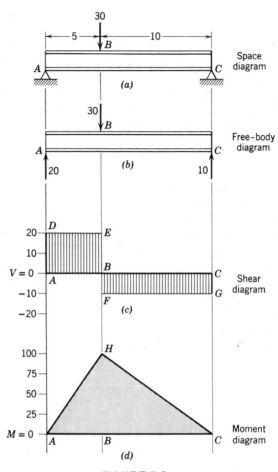

FIGURE 7-2

STEP 1.

Draw the free-body diagram of the beam.
The reactions of the supports at A and C are found by summing moments about A and C respectively. The space scale of the diagram should be reasonably accurate. Complete as in Figure 7-2(b).

STEP 2.

Draw the shear diagram of the beam.
Drop fine guidelines down from points A, B, and C of the free-body diagram.

Then establish the zero-shear line ($V = 0$) at any convenient location and parallel with the beam in the free-body diagram.

Remembering that the shear at any point is equal to the sum of the forces and reactions to the left of that point, we begin by considering the point immediately to the right of point A. The only force to the left of this point is the reaction at A which is 20 units acting upward. Therefore, a line is drawn from point A of the $V = 0$ line vertically upward 20 units to some convenient scale to point D. As we proceed along the beam toward the right no other force acts on the beam until we come to point B, so the shear force remains constant for that portion of the beam and we may draw the line from D to E on the shear diagram. As we pass point B, where the 30 unit concentrated load acts, the sum of our shearing forces becomes $20 - 30 = -10$ so we draw the line EBF vertically downward to -10. Again as we move toward the right there are no new forces acting on the beam and the shear remains constant as shown by line FG. At the right-hand end of the beam the 10 unit upward-acting reaction brings our diagram back to $V = 0$ at point C, thus completing the shear diagram.

STEP 3.

Draw the Bending Moment Diagram.

Here again we drop guidelines down from the above points A, B, and C. At a convenient location we draw the $M = 0$ line parallel to the beam of the free-body diagram.

The bending moment at any point of the beam is equal to the sum of the moments of the forces to the left of that point. Since there are no forces to the left of point A, the bending moment at A is zero, which is the starting point of our diagram. As we proceed to the right toward B the bending moment at any point is the sum of the moments of the forces to the left of that point, in this case the moment of the single 20-unit left-hand reaction. The bending moment at any point between A and B is

$$M = 20d$$

where d is the distance of the point from A. This is a straight-line relationship so we need only calculate the moment at point B which is $20(5) = 100$ units. Thus our moment diagram starts out as a diagonal line from zero at A to 100 at H. As we move toward the right past point B the negative moment caused by the downward acting 30 unit concentrated load is subtracted from the moment of the left-hand 20 unit reaction. Thus at any point to the right of point B the bending moment is

$$M = 20d - 30(d - 5)$$

Again this is a straight line mathematical relationship which gives us the straight line from H to C as shown in Figure 7-2(d). The shading shown on the shear and bending moment diagrams is optional. Some people like to shade the areas for emphasis.

Example of a simple beam with uniform loading

Figure 7-3(a) shows the simply supported beam AB carrying a uniform load. To simplify the example we will ignore the weight of the beam itself, although in many design situations it may be significant and should be considered.

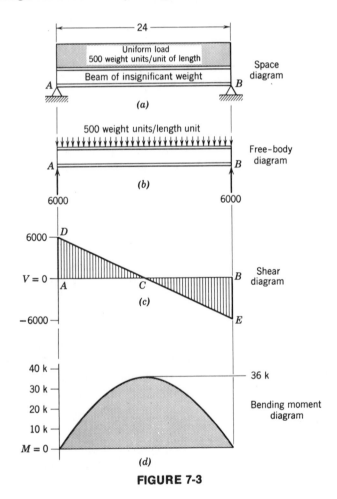

FIGURE 7-3

STEP 1.

Draw the free-body diagram as shown in Figure 7-3(b).
In this case the uniform load covers the entire length of the beam, giving us symmetrical loading. The total uniform load ($500 \times 24 = 12\,000$) will divide equally between the supports giving the 6 000 unit reactions at each end as shown.

STEP 2.

Draw the shear diagram as shown in Figure 7-3(c).
We start our diagram by drawing line AD vertically upward from A to show the effect of the 6 000 unit left-hand reaction. Due to the effect of the downward force of the uniform load, the upward shear is reduced by 500 units for each foot or meter we move toward the right from A to B along the beam. This provides us with the straight line DCE shown. At point E the shear, which has reached a value of $-6\,000$ units, is brought back up to 0 at point B by the effect of the right-hand reaction.

STEP 3.

Draw the bending moment diagram as shown in Figure 7-3(d).
As previously stated the bending moment at any point is equal to the sum of the moments of all of the loads and reactions to the left of that point. In calculating the moment of the uniform load to the left of any point on the beam, the entire weight of the portion of the uniform load to the left of the point is considered as concentrated at the center of that portion of the uniform load and is dealt with as a concentrated load acting at that point. Calculating the bending moment for various points on the beam and plotting them will define a parabolic curve as shown, with the maximum moment being at the center where the shear is zero.

The procedure of summing the moments in Step 3 above can be quite tedious. It is here, however, that two simple relationships between the shear and bending moment diagrams come in handy. *The bending moment for any point on the beam is equal to the area of the shear diagram to the left (or right) of that point*, taking areas above the $V = 0$ line to be positive and those below to be negative. Also, *the slope of the moment diagram at any point is proportional to the shear at that point*. For instance, if the shear is constant for a portion of the shear diagram, the bending moment diagram for that portion will have a constant slope, that is, it will be a straight line. Similarly, if the shear is changing at a uniform rate, the slope of the bending moment diagram will be changing at a uniform rate. We will now use the information developed thus far to construct the shear and bending moment diagrams for a beam with a rather complex loading and support arrangement.

Example of an overhanging beam supporting both concentrated and uniform loads

We are to construct the shear and bending moment diagrams for the beam shown in Figure 7-4.

STEP 1.

Here we draw the free-body diagram as in the previous examples. We will simplify our drawing, however, by letting a single line represent the beam and by indicating the uniform load as shown in Figure 7-4(b). The reaction R_E is determined by summing moments about A. For determining the reactions the entire uniform load (10 times 200 units or 2 000 units) is considered to be concentrated at its center which, in this example, is one length unit to the right of E.

$$\sum M_A = 0$$

$$\sum M_A = 1(800) + 4(600) + 11(2\,000) - 10R_E = 0$$

$$R_E = 2520$$

Summing the vertical forces then establishes R_A to be 880.

STEP 2.

We can now construct the shear diagram from the information on the free-body diagram. We start at A by going perpendicularly upward 880 units, the magnitude of R_A, to A'. There being no forces acting between A and B, we draw the horizontal line $A'B'$. At B the 800 unit force brings our shear down to 80 where it remains until we reach C. This is shown by the horizontal line from B to C. At C the 600 unit force dictates the vertical line from C to C', giving us a shear value of -520 units at C'. Nothing happens between C and D so we draw the horizontal line $C'D'$. Beginning at D we encounter the uniform load which increases the shear in the negative direction at the rate of 200 units per foot or meter, giving us the sloping line $D'E''$ with E'' at -1320 units. At this point R_E boosts us up 2520 units to $+1200$ where the uniform load again takes over, bringing the shear down to zero at F as shown.

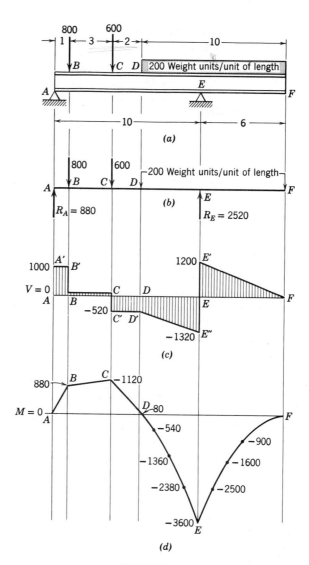

FIGURE 7-4

171

STEP 3.

We can now construct the bending moment diagram by analyzing the areas of the shear diagram. We begin by observing that the area to the left of point B is $(1)(880) = 880$ units. Also the shear is constant from A' to B' of the shear diagram, indicating that the bending moment diagram varies at a constant rate thus giving us a straight line from A to B as seen in Figure 7-4(d). Adding the $(3)(80) = 240$ unit area between B and C to the previously determined 880 gives us a positive bending moment value of 1 120 units at C with the line between B and C being straight. The area of the shear diagram between C and D is negative so we subtract its value (1 040 units) from the previous 1 120 to find the value of the bending moment at D to be $+80$. Between D and E the shear is increasing in the negative direction at a constant rate, which tells us that the slope of the bending moment diagram will be changing at a constant rate also. The bending moment diagram between D and E will be a curved line. To plot it we will have to calculate the value of the bending moment at a sufficient number of points to enable us to draw a smooth and accurate curve. To determine the points on the curve between D and E we sum the areas of the shear diagram to the left of a point. As the diagram passes point E the shear again becomes positive and the area to the right of E in the shear diagram is positive, again sloping at a constant rate. The moment diagram therefore proceeds from E to F along the curve as shown in Figure 7-4(d) and ends at $M = 0$ at F.

As previously mentioned the maximum moment will occur at a point where the shear on the beam is zero. We can see that there are four points where the shear is zero—at A, C, E, and F. It can be seen at a glance that the maximum moment is at E and has a value of 3600 units (lb-ft or $N \cdot m$).

Example of a cantilever beam

Figures 7-5 and 7-6 show the development of the shear and bending moment diagrams for a cantilever beam. The procedure is the same as that in the previous examples, however, there are some points worth noting. In drawing the free-body diagram we must realize that the entire support for the beam must be provided at the point where the beam becomes free of its rigid mounting. Thus the reaction R must balance the sum of all of the loads acting downward, while the couple (moment) M must balance the moments of all of those loads about point D. Thus the maximum shear will occur immediately to the left of point D and the maximum bending moment will occur at point D where the shear is zero.

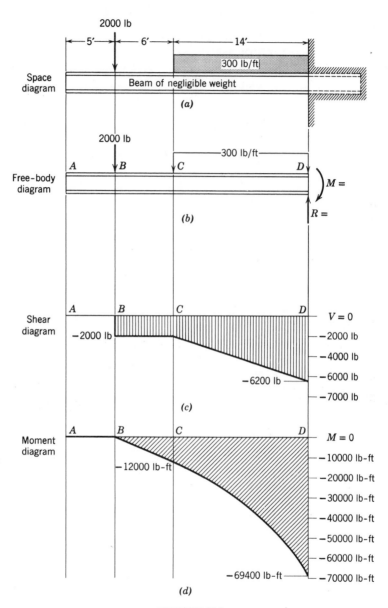

FIGURE 7-5

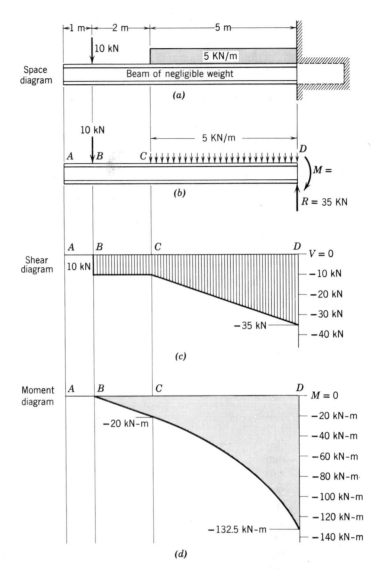

FIGURE 7-6

7-4 ACCOMPLISHMENT CHECKLIST

From your study of this chapter you should now understand

1. The difference between homogeneous beams, hybrid beams, composite beams, and girders.

2. The difference between statically determinate and statically indeterminate beams.

3. The difference between a simple beam, cantilever beam, overhanging beam, and a fixed beam.

4. The terms *shearing force* and *bending moment.*

5. How to draw shear and bending moment diagrams.

6. How to use shear and bending moment diagrams to find the critical sections of beams where the maximum stresses will occur.

7. The relations between load intensity, shearing force, and bending moment.

7-5 SUMMARY

A *beam* is a member of a structure or machine which is designed to sustain loads acting perpendicular to its axis. *Homogeneous beams* have uniform material throughout. *Hybrid beams* have material of higher strength in the flanges than in the web. *Composite beams* are made of more than one piece of different materials inseparably joined to act as a unit. Large beams fabricated from plates may be further classified as girders.

Beams which can be analyzed by the methods of statics are *statically determinate*, otherwise they are *statically indeterminate*.

A *simple beam* is supported at its ends in such a manner that expansion or flexure is not restricted. If one or more ends of a simply supported beam extend beyond a support, they are called *overhanging beams*. A *cantilever beam* has one end rigidly imbedded in a support, while the other end hangs free. A *fixed beam* has both ends rigidly supported, thus preventing their movement.

The *shearing force* on the cross section of a beam is the algebraic sum of the applied loads and reaction on either side of the cross section. The *bending moment* of any beam at any cross section perpendicular to the longitudinal axis of a beam is the sum of the moments of the loads and reactions on either side of that cross section.

PROBLEMS

7-1C   Draw the shear and moment diagrams for the three beams. Determine the maximum shear and bending moment and locate the critical section of each beam. *P* is 20 kips, *a* is 15 ft, *b* is 5 ft, and *L* is 20 ft. The uniform load is 500 lb/ft.

PROBLEM 7-1

7-1M   *Draw the shear and moment diagrams for the three beams. Determine the maximum shear and bending moment and locate the critical section of each beam. P is 20 kN, a is 10 m, b is 5 m, and L is 15 m. The uniform load is 750 kg/m.*

7-2C   Draw the shear and moment diagrams for the three beams. Determine the maximum shear and bending moment and locate the critical section or sections for each beam. *P* is 6000 lb, *a* and *c* are 5 ft, *b* is 8 ft, *d* is 10 ft, *L* is 18 ft, and the uniform load is 300 lb/ft.

PROBLEM 7-2

7-2M *Draw the shear and moment diagrams for the three beams. Determine the maximum shear and bending moment and locate the critical section or sections for each beam. P is 25 kN, a and c are 2 m, b is 3 m, d is 4 m, L is 7 m, and the uniform load is 450 kg/m.*

7-3C Draw the shear and bending moment diagrams for the overhanging beam. Determine the critical section. P_1 is 1000 lb, P_2 is 500 lb, a is 5 ft, d is 15 ft, and L is 20 ft.

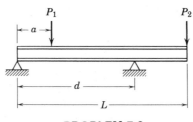

PROBLEM 7-3

7-3M *Draw the shear and bending moment diagrams for the overhanging beam and determine the critical section. P_1 is 4 kN, P_2 is 2 kN, a is 2 m, d is 6 m, and L is 8 m.*

7-4C The cantilever beam shown supports a uniform and a concentrated load. Draw the shear and bending moment diagrams and determine the bending moment at the critical section. P is 2000 lb, L is 20 ft, d is 15 ft, and the uniform load is 300 lb/ft.

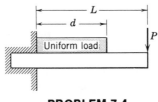

PROBLEM 7-4

7-4M *The cantilever beam shown supports a uniform load and a concentrated load. Draw the shear and bending moment diagrams and locate the critical section. P is 8 kN, L is 6 m, d is 4 m, and the uniform load is 450 kg/m.*

7-5C The figure shows a simple beam carrying a uniform load and a concentrated load. Determine the critical section or sections by drawing the shear and bending moment diagrams. P is 4000 lb, a is 10 ft, b is 6 ft, c is 4 ft, and the uniform load is 500 lb/ft.

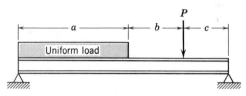

PROBLEM 7-5

7-5M *The figure shows a simple beam carrying a uniform load and a concentrated load. Determine the critical section or sections by drawing the shear and bending moment diagrams. P is 18 kN, a is 3 m, b is 2 m, c is 2 m, and the uniform load is 750 kg/m.*

7-6C The cantilever beam shown is made of oak wood and weighs 80 lb per lineal foot. The suspended load W weighs 1000 lb, the uniform load weighs 200 lb/ft, a is 10 ft, and L is 15 ft. Draw the shear and bending moment diagrams.

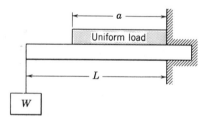

PROBLEM 7-6

7-6M *The cantilever beam shown is made of oak wood and weighs 120 kg per lineal meter. The suspended load W weighs 500 kg, the uniform load weighs 300 kg/m, a is 4 m, and L is 6 m. Draw the shear and bending moment diagrams.*

7-7C Draw the shear and bending moment diagrams for the beam shown which overhangs both supports and carries a uniform load of 600 lb/ft. L is 50 ft, a is 8 ft, b is 6 ft.

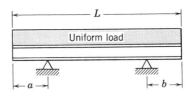

PROBLEM 7-7

7-7M   *Draw the shear and bending moment diagrams for the beam shown which overhangs both supports and carries a uniform load of 900 kg/m. L is 15 m, a is 2.5 m, and b is 2 m.*

7-8C   The figure shows the cross section of a loading dock which is supported on steel beams which weigh 260 lb/ft. The concrete deck portion per beam weighs 700 lb per lineal foot. The load intensity of the storage bin is 5000 lb/ft and the bricks supported by this beam weigh 5 tons. *L* is 85 ft, *a* is 40 ft, *b* is 20 ft, *c* is 10 ft, *d* is 10 ft, and *e* is 10 ft. Draw the shear and bending moment diagrams and locate the critical section or sections of the beam.

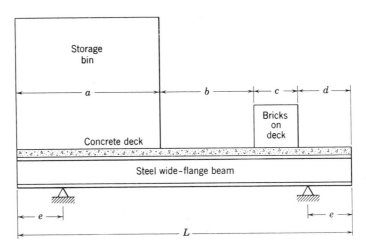

PROBLEM 7-8

7-8M   *The figure shows the cross section of a loading dock which is supported on steel beams which weigh 390 kg/m. The concrete deck portion per beam weighs 1 000 kg per lineal meter. The load intensity of the storage bin is 7 500 kg/m and the bricks supported by this beam weigh 5 metric* tons. *L is 24 m, a is 12 m, b is 6 m, c, d, and e are 3 m. Draw the shear and bending moment diagrams and locate the critical section or sections of the beam.*

7-9C   The cantilever beam shown has a uniform load of 100 lb/ft and concentrated loads $P_1 = 400$ lb and $P_2 = 800$ lb. L is 15 ft, a is 11 ft, and d is 8 ft. Draw the shear and bending moment diagrams assuming that the beam's weight is not significant.

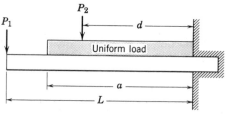

PROBLEM 7-9

7-9M   *The cantilever beam shown has a uniform load of 150 kg/m and concentrated loads $P_1 = 200$ kg and $P_2 = 400$ kg. L is 5 m, a is 3 m, and d is 2.5 m. Draw the shear and bending moment diagrams assuming that the beam's weight is not significant.*

7-10C   The figure shows a beam weighing 80 lb/ft carrying two uniform loads. Uniform load 1 is 200 lb/ft, uniform load 2 is 400 lb/ft, a is 12 ft, b is 13 ft, and L is 32 ft. Draw the shear and bending moment diagrams for the beam and determine the critical section.

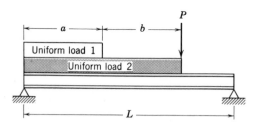

PROBLEM 7-10

7-10M   *The figure shows a beam weighing 120 kg/m carrying two uniform loads. Uniform load 1 is 300 kg/m, uniform load 2 is 600 kg/m, a is 4 m, b is 5 m,* *and L is 12 m. Draw the shear and bending moment diagrams for the beam and determine the critical section.*

7-11C   Draw the shear and bending moment diagrams for the beam and concrete slab when the weight of the uniform load and concrete slab together is 400 lb/ft and the beam weighs 60 lb/ft. L_1 is 20 ft, L_2 is 10 ft, L_3 is 15 ft, and L_4 is 1 ft.

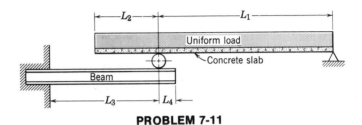

PROBLEM 7-11

7-11M   *Draw the shear and bending moment diagrams for the beam and concrete slab when the weight of the uniform load and concrete slab together is 600 kg/m and the beam weighs 90 kg/m. L_1 is 8 m, L_2 is 4 m, L_3 is 6 m, and L_4 is 0.5 m.*

7-12C   The simple beam in the figure carries a load that varies uniformly from zero at B to 500 lb per ft at C. Draw the shear and moment diagrams and locate the critical section. The beam's weight shall be neglected. L is 10 ft, and a is 6 ft.

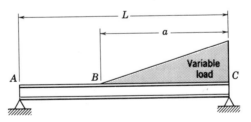

PROBLEM 7-12

7-12M   *The simple beam in the figure carries a load that varies uniformly from zero at B to 255 kg per meter at C. Draw the shear and moment diagrams and locate the critical section. L is 10 m, a is 6 m. The beam's weight shall be neglected.*

SHEAR FORCE AND BENDING MOMENT IN BEAMS

7-13C   The cantilever beam shown carries a load that varies from zero at *A* to 800 lb per ft at *B*. Draw the shear and bending moment diagrams assuming the weight of the beam to be negligible. *L* is 15 ft.

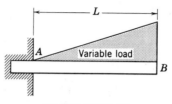

PROBLEM 7-13

7-13M   *The cantilever beam shown carries a load that varies from zero at A to 408 kg per meter at B. Draw the shear and bending moment diagrams assuming the weight of the beam to be negligible. L is 5 m.*

7-14C   The overhanging beam in the figure carries a load that varies from 1000 lb/ft, at *A* to zero at *B*. *L* is 20 ft, and *d* is 15 ft. Draw the shear and bending moment diagrams and locate the critical section of the beam.

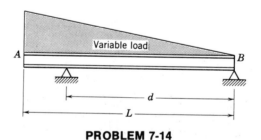

PROBLEM 7-14

7-14M   *The overhanging beam in the figure carries a load that varies from 510 kg/m at A to zero at B. L is 8 m, d is 6 m. Draw the shear and bending moment diagrams and locate the critical section of the beam.*

7-15C   Draw the shear and bending moment diagrams for the cantilever beam shown if F_1 is 300 lb/ft, F_2 is 600 lb/ft, F_3 is 900 lb/ft, *a* is 2 ft, *b* is 12 ft, *c* is 6 ft, and *d* is 4 ft. The weight of the beam is negligible.

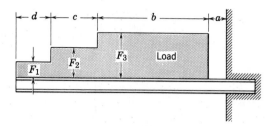

PROBLEM 7-15

7-15M   Draw the shear and bending moment diagrams for the cantilever beam shown if F_1 is 306 kg/m, F_2 is 612 kg/m, F_3 is 936 kg/m, a is 1 m, b is 6 m, c is 3 m, and d is 2 m. The weight of the beam is negligible.

CHAPTER
OBJECTIVES ————————————————

By the study of this chapter you will learn how to

1. Determine the normal bending stresses in statically determinate homogeneous beams of uniform cross section.

2. Determine the section moduli of the above beams.

3. Use section moduli to select the most economical size and shape of beam for the job.

4. Determine the shearing stresses in beams.

5. Determine the bending and shearing stresses in beams of two materials.

CHAPTER ———————————— 8
STRESSES IN BEAMS

Having learned how to determine the maximum shear forces and bending moments on beams under various loadings in the previous chapter, we are now ready to determine the stresses caused by them. We are going to examine some of the fundamentals of beam design. The ability to determine the stresses in beams will enable us to select the proper beams for given load conditions. Because the design of some beams can be very complex, we will confine our studies to the more straightforward types, which account for most of the beams used. Our first thought at the mention of beams may be of beams used in structures such as buildings and bridges. It is important to note, however, that many members of mechanisms and machines such as levers and shafts—even rotating shafts—are acting as beams with respect to the transverse loads imposed on them.

8-1 BENDING STRESSES IN HOMOGENEOUS BEAMS

In this section we will concern ourselves only with beams on which all loads lie in one plane and act perpendicular to the centroidal axis of the beam. Beams under such loading are not subject to bending in more than one plane or torsion. Also, our studies will be concerned with beams that have sufficient lateral support to prevent buckling. Our analysis will be based on beams in which all stresses are within the elastic range, that is, below the yield stress.

   Figure 8-1 shows a simple beam acted on by two concentrated loads P located equidistant from either end of the beam. As can be seen in the shear

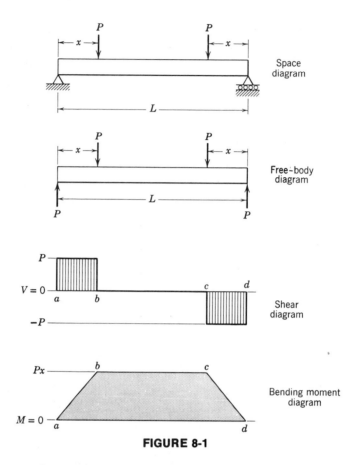

FIGURE 8-1

diagram the vertical shear between b and c is zero, while the bending moment as shown on the bending moment diagram is constant between b and c. Thus between b and c we have *pure bending*. This condition seldom occurs in beams but it will serve us well in analyzing the stress situation.

Figure 8-2 is a representation of the beam as it might actually appear under load. As the load is applied the beam bends to the position shown. Part (b) of the figure is the free-body diagram of the left half of the beam with the right half removed. Note that the moment M provides the entire support of the beam at its center. If there were other than pure bending in this section of the beam, there would have to be a shearing force at the cut in addition to the bending moment.

The bending moment of part (b) of Figure 8-2 is supplied by the forces on each particle of the cut section of the beam acting as indicated in part (c) of

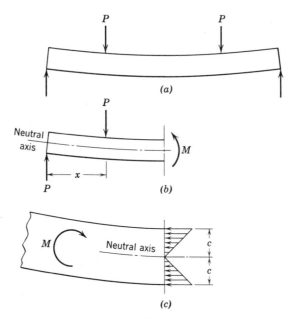

FIGURE 8-2

the figure. As the beam bends the particles of the top portion of the beam are being compressed as shown by the arrows indicating forces pushing on the section. Similarly, we can see that the particles of the lower portion of the beam, being pulled on, are in tension. It follows from this that the top of the beam becomes shorter under load due to compression, while the bottom of the beam is stretched and becomes longer. Somewhere in between there must be a place where there is no change in the length of its fibers. This happens on a plane perpendicular to the direction of the applied loads called the *neutral plane,* the longitudinal centerline of which is the *neutral axis.* With a homogeneous material the stress on any particle is proportional to its distance above or below the neutral plane. The neutral axis is at the centroid of the cross section of the beam. Let f_b be the stress due to bending of the beam on the particle furthest from the neutral axis of the beam, c be the distance from the neutral plane to that particle, and f_p be the stress on any particle at distance y from the neutral axis. All of these particles are to be visualized as being infinitely small with an area in the plane of the cut being ΔA. From the proportionality of stress to the distances from the neutral axis we can write the equality

$$\frac{f_b}{c} = \frac{f_p}{y} \qquad (8\text{-}1)$$

187

The bending moment M of the beam must be balanced by the sum of the moments of all forces on all particles of the cut section of the beam. Mathematically, this means

$$M = \sum (f_p \Delta A) y$$

From Equation 8-1, $f_p = f_b y/c$. Substituting this value of f_p in the above equation gives us

$$M = \frac{\sum f_b y^2 \Delta A}{c}$$

The term $\sum y^2 \Delta A$ of the above equation is mathematically the moment of inertia I of the area of the cross section. Therefore we can write $M = f_b I/y$, or when solved for f_b,

$$f_b = \frac{My}{I} \tag{8-2}$$

For the particles furthest from the neutral axis, the *extreme fibers*, Equation 8-2 becomes

$$f_b = \frac{Mc}{I} = \frac{M}{S} \tag{8-3}$$

where S, ($= I/c$), is the *elastic section modulus* of the beam. The subscripts x and y are used with S to designate the neutral plane being bent. Although Equation 8-3 was developed for a beam in perfect bending, it may be used to determine f_b, *the bending stress*, of any beam of homogeneous material loaded so that there is no tendency for the beam to bend in more than one direction or twist, with it being sufficiently supported to prevent buckling. The examples that follow demonstrate the application of Equation 8-3 in determining the maximum stress, first in a wide-flange beam which is symmetrical in two perpendicular directions, then in a T beam which is symmetrical only about the axis of loading.

Example Using Wide-Flange Beam

customary

Figure 8-3 shows a W10X19 wide-flange beams' free-body, shear, and bending moment diagrams. Determine the maximum tensile and compressive bending stresses in the beam.

metric

Figure 8-4 shows a 250 × 125 − 29.6 wide-flange beam's free-body, shear, and bending moment diagrams. Determine the maximum tensile and compressive bending stresses in the beam.

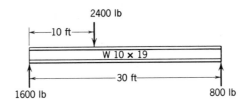

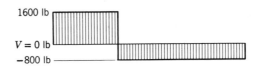

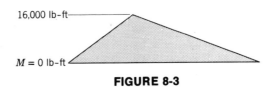

FIGURE 8-3

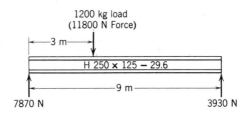

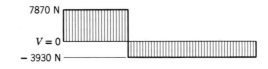

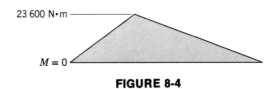

FIGURE 8-4

STRESSES IN BEAMS

Solution

customary

Since wide-flange beams are symmetrical with respect to both the vertical and horizontal axes, the neutral axis will be at the geometrical center of the cross section and the distance to the extreme fiber will be the same for compression and tension. It follows that the maximum stress in tension will equal the maximum stress in compression.

From Appendix 2 we find S_x, the section modulus with respect to the X-X axis about which the beam is bending, to be 18.8 in.³ From Figure 8-3 we find the maximum bending moment to be 16,000 lb-ft. Using these values we determine the maximum bending stress to be

$$f_b = \frac{M}{S_x} = \frac{16,000 \text{ lb-ft}(12 \text{ in./ft})}{18.8 \text{ in.}^3}$$
$$= 10,200 \text{ psi}$$

metric

Since wide-flange beams are symmetrical with respect to both the vertical and horizontal axes, the neutral axis will be at the geometrical center of the cross section and the distance to the extreme fiber will be the same for compression and tension. It follows that the maximum stress in tension will equal the maximum stress in compression.

From Appendix 2 we find S_x, the section modulus with respect to the X-X axis about which the beam is bending, to be 324 cm³ which is $3.24 \times 10^{-4} m^3$. From Figure 8-4 we find the maximum bending moment to be 23 600 N·m Using these values we determine the maximum bending stress to be

$$f_b = \frac{M}{S_x} = \frac{23\,600 \text{ N·m}}{3.24 \times 10^{-4} m^3} = 72.8 \text{ MPa}$$

Example Using a Tee Beam

customary

Figure 8-5(a), (b), and (c) shows the free-body, shear, and bending moment diagrams of a Tee beam of cross section shown to the right of part (a) of the figure. Determine the maximum tensile and compressive stresses in the beam.

metric

Figure 8-6(a), (b), and (c) shows the free-body, shear, and bending moment diagrams of a Tee beam of cross section shown to the right of part (a) of the figure. Determine the maximum tensile and compressive stresses in the beam.

Solution

customary

Because this is a fabricated Tee and we have no table available to list its proper-

metric

Because this is a fabricated Tee and we have no table of values available,

190

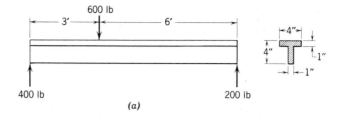

(a)

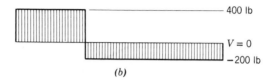

(b)

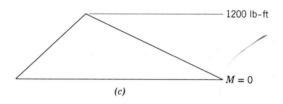

(c)

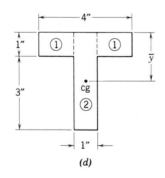

(d)

FIGURE 8-5

191

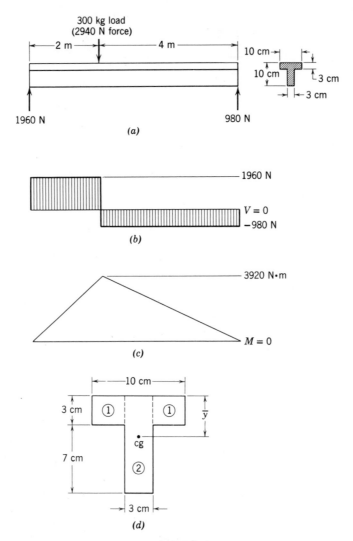

FIGURE 8-6

ties, we will have to determine the location of the neutral axis (at its centroid), its moment of inertia, and its elastic section modulus.

Referring to Figure 8-5(d)

we will have to determine the location of the neutral axis (at the centroid of the cross section), its moment of inertia, and its elastic section modulus.

Referring to Figure 8-6(d)

192

$$\bar{y} = \frac{A_1 y_1 + A_2 y_2}{A_1 + A_2} = \frac{3(0.5) + 4(2)}{3 + 4} = 1.36 \text{ in.}$$

Again referring to Figure 8-5(d) we can determine the moment of inertia of the cross section about the neutral axis as follows.

$$I_1 = \frac{bh^3}{12} + Ad^2$$

$$= \frac{3(1)^3}{12} + 3(0.5 - 1.36)^2 = 2.47 \text{ in.}^4$$

$$I_2 = \frac{bh^3}{12} + Ad^2$$

$$= \frac{1(4)^3}{12} + 4(2 - 1.36)^2 = 6.97 \text{ in.}^4$$

$$I = 9.44 \text{ in.}^4$$

We can now find S.

For the top of the beam which is in compression, $c = 1.36$ in.

$$S_{top} = \frac{I}{c} = \frac{9.44 \text{ in.}^4}{1.36 \text{ in.}} = 6.94 \text{ in.}^3$$

The maximum compressive stress will be

$$f_b = \frac{M}{S} = \frac{1200 \text{ lb-ft}(12 \text{ in./ft})}{6.94 \text{ in.}^3} = 2070 \text{ psi}$$

For the bottom of the beam which is in tension

$$c = 4.00 \text{ in.} - 1.36 \text{ in.} = 2.64 \text{ in.}$$

$$S_{bottom} = \frac{I}{c} = \frac{9.44 \text{ in.}^4}{2.64 \text{ in.}} = 3.58 \text{ in.}^3$$

The maximum tensile stress will be

$$f_b = \frac{M}{S} = \frac{1200 \text{ lb-ft}(12 \text{ in./ft})}{3.58 \text{ in.}^3} = 4020 \text{ psi}$$

$$\bar{y} = \frac{A_1 y_1 + A_2 y_2}{A_1 + A_2}$$

$$= \frac{21 \text{ cm}^2(1.5 \text{ cm}) + 30 \text{ cm}^2(5 \text{ cm})}{21 \text{ cm}^2 + 30 \text{ cm}^2}$$

$$= 3.56 \text{ cm}$$

Again referring to Figure 8-5(d) we can determine the moment of inertia of the cross section about the neutral axis as follows.

$$I_1 = \frac{bh^3}{12} + Ad^2 = \frac{7 \text{ cm }(3 \text{ cm})^3}{12}$$

$$+ 21 \text{ cm}^2(3.56 \text{ cm} - 1.5 \text{ cm})^2 = 105 \text{ cm}^4$$

$$I_2 = \frac{bh^3}{12} + Ad^2 = \frac{3 \text{ cm }(10 \text{ cm})^3}{12}$$

$$+ 30 \text{ cm}^2(3.56 \text{ cm} - 5 \text{ cm})^2 = 312 \text{ cm}^4$$

$$I = I_1 + I_2 = 105 \text{ cm}^4 + 312 \text{ cm}^4$$

$$= 417 \text{ cm}^4$$

For the top of the beam—in compression $c = 3.65$ cm

$$S_{top} = \frac{I}{c} = \frac{417 \text{ cm}^4}{3.56 \text{ cm}} = 117 \text{ cm}^3$$

$$= 117 \times 10^{-6} \text{ m}^3$$

The maximum compressive stress will be

$$f_b = \frac{M}{S} = \frac{3920 \text{ N·m}}{117 \times 10^{-6} \text{ m}^3} = 33.5 \text{ MPa}$$

The bottom of the beam is in tension. For it

$$c = 10 \text{ cm} - 3.56 \text{ cm} = 6.44 \text{ cm}$$

$$S_{bottom} = \frac{I}{c} = \frac{417 \text{ cm}^4}{6.44 \text{ cm}} = 64.8 \text{ cm}^3$$

$$= 64.8 \times 10^{-6} \text{ m}^3$$

The maximum tensile stress will be

$$f_b = \frac{M}{S} = \frac{3920 \text{ N·m}}{64.8 \times 10^{-6} \text{ m}^3} = 60.5 \text{ MPa}$$

The more common occurrence in practice is the situation where a beam is required to span a given distance with a given loading pattern. The designer must then select the beam that will do the job. In such a situation you can determine the maximum bending moment and thus select the proper grade of steel for the job. Having selected the steel you then have available its allowable bending strength F_a. Knowing the bending moment and the bending strength you can use Equation 8-3 to determine the required elastic section modulus. When you find the section modulus, search the tables of design properties, Appendix 2, for the lowest weight of beam of the type suitable for the application that has a section modulus equal to or greater than the one required. Because steel is basically priced on a tonnage basis, this will be the most economical beam.

Example

customary

You find the maximum bending moment on a beam to be 2.88×10^6 lb-in. It is decided to use a wide-flanged beam made of A36 steel. The allowable bending stress is 24,000 psi. Determine the most economical beam among those listed in Appendix 2.

metric

You find the maximum bending moment on a beam to be 325 kN·m. It is decided to use a wide-flanged beam made of steel having an allowable bending stress of 165 MPa. Determine the most economical beam among those listed in Appendix 2.

Solution

customary

$$S = \frac{M}{F_a} = \frac{2.88 \times 10^6 \text{ lb-in.}}{24,000 \text{ psi}} = 120 \text{ in.}^3$$

In Appendix 2 you find the lowest weight beam having an S of no less than 120 in.^{3} to be a W24X61.

metric

$$S = \frac{M}{F_a} = \frac{325 \text{ kN·m}}{165 \text{ MPa}} = 1.97 \times 10^{-3} \text{ m}^3$$
$$= 1970 \text{ cm}^3$$

In Appendix 2 you find the lowest weight beam having an S of no less than 1970 cm³ to be 600 × 200 − 94.6.

8-2 HORIZONTAL AND VERTICAL SHEARING STRESSES

In Figure 8-7(a) we have a rectangular steel beam (or beam of any other elastic material which obeys Hooke's law) mounted as a simple beam with rollers at each of its ends. In part (b) of the figure the beam has been slit parallel to its neutral surface on the plane CD, thus making it into two beams. When the load P is placed on it as in part (c) of the figure the two beams bend, taking the position shown. In actual loading practice the bend would be but a fraction of that shown in the figure. As the beams bend the bottom fibers become longer as demonstrated by the spreading of the supports. Similarly, due to the compressive forces at the tops of the beams the tops become equivalently shorter. Obviously there has been a sliding of the top beam over the bottom beam along the CD plane. Now if the load P were removed, letting the beams go back to the unloaded position and they were magically reunited into one beam, the replacing of the load P would cause the beam to

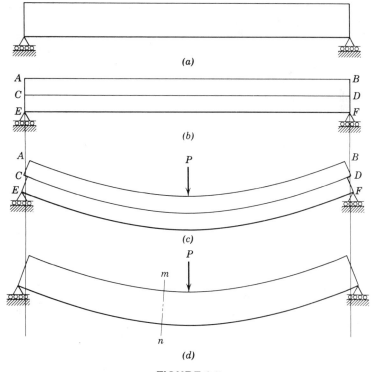

FIGURE 8-7

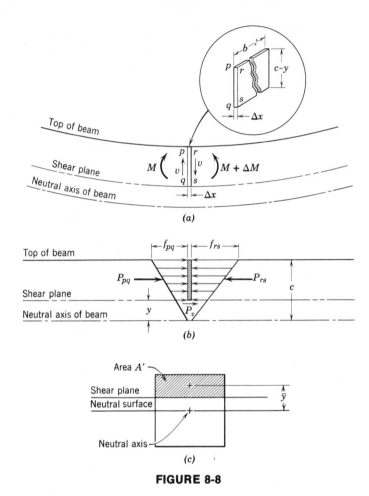

FIGURE 8-8

assume the position shown in part (d) of the figure. In this case, however, the natural tendency for the bottom of the top portion of the beam to slide over the top of the bottom half is prevented. Horizontal shearing of the beam has been prevented by a force—THE HORIZONTAL SHEARING FORCE.

To develop an equation that may be used to determine the horizontal shearing stress we will examine a very thin section taken from the vicinity of plane mn of the beam in Figure 8-7(d). The free-body diagram of the section is Figure 8-8(a). It extends from the shear plane to the top of the beam and includes the full width of the beam as shown in the pictorial insert. Note that the moment on the right side of the section is larger than that on the left by ΔM.

Figure 8-8(b) is a free-body diagram showing the horizontal forces and stresses acting at the surfaces of the section. From Equation 8-2 the forces within the triangular stress patterns are

$$f_{pq} = \frac{My}{I} \quad \text{and} \quad f_{rs} = \frac{(M + \Delta M)y}{I}$$

P_v is the shear force at the bottom (shear surface) of the section. P_{pq} is the resultant of the compressive stress acting on the pq side of the section and P_{rs} is that on the rs side. Due to the larger moment on the rs side, P_{rs} is larger than P_{pq}.

$$P_{pq} = \sum f_{pq}\Delta A \quad \text{and} \quad P_{rs} = \sum f_{rs}\Delta A$$

where ΔA represents a miniscule portion of the surface of the section, each ΔA portion being acted on by the correct f_{pq} or f_{rs} for its location. The sum of all the ΔAs on a side of the section equals the total area A' depicted in Figure 8-8(c).

In order for the section as shown in Figure 8-8(b) to be in equilibrium

$$P_v = P_r - P_{pq}$$

Substituting the previously determined values of P_{pq} and P_{rs} in the above equation yields

$$P_v = \sum \frac{(M + \Delta M)y}{I}\Delta A - \sum \frac{My}{I}A$$

which reduces to

$$P_v = \sum \frac{\Delta M}{I}y\Delta A$$

Since $P_v = f_v A = f_v b\Delta x$ we can write the above equation

$$f_v b\Delta x = \sum \frac{\Delta M}{I}y\Delta A$$

Solving for f_v gives us

$$f_v = \sum \frac{\Delta M}{I}\frac{y\Delta A}{b\Delta x}$$

Since ΔM, Δx, I, and b are constants this may be written

$$f_v = \frac{\Delta M}{x}\frac{\sum y\Delta A}{Ib} \tag{8-4a}$$

By summing moments in Figure 8-8(a) we find

$$M + V\Delta x = M + \Delta M$$

which yields

$$V = \frac{\Delta x}{\Delta M}$$

Referring to Figure 8-8(c) we can determine the distance from the neutral axis to the centroid of area A' to be

$$\bar{y} = \frac{\Sigma \, y\Delta A}{A'}$$

or

$$\sum y\Delta A = A'\bar{y}$$

Substituting the above values of V and $\Sigma \, y\Delta A$ in Equation 8-4(a) produces

$$f_v = \frac{VA'\bar{y}}{Ib} \qquad\qquad (8\text{-}4)$$

where f_v is the shearing stress, V is the vertical shearing force, A' is the area of that portion of the cross section lying beyond the shear plane with respect to the neutral axis, $\bar{y}$ is the distance from the neutral axis to the centroid of A', I is the moment of inertia of the entire cross section of the beam, and b is the width of the beam at the shear plane. *The maximum shearing stress will occur on the neutral plane.*

You may have noted that f_v was referred to above as the shearing stress, not the horizontal shearing stress as might have been expected. The reason is that it has been established that the vertical shearing stress is equal to the horizontal shearing stress. Thus we use f_v to represent either.

Example

customary

Figure 8-9 shows the cross section of an aluminum box girder of which b is 3 in., h is 5 in., and t is 0.5 in. Determine the maximum shearing stress in the beam if it is 6 ft long, is simply supported, and

metric

Figure 8-9 shows the cross section of an aluminum box girder of which b is 7 cm, h is 12 cm, and t is 1 cm. Determine the maximum shearing stress in the beam if it is 2 m long, is simply

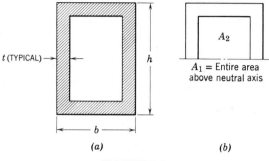

FIGURE 8-9

supports an 8000 lb load at the middle of its span.

supported, and supports a load of 4000 kg at the middle of its span.

Solution

customary

$$V = \frac{8000 \text{ lb}}{2} = 4000 \text{ lb}$$

$$I = \frac{3 \text{ in. } (5 \text{ in.})^3}{12} - \frac{2 \text{ in. } (4 \text{ in.})^3}{12} = 20.6 \text{ in.}^4$$

$$t = 2 \times 0.5 \text{ in.} = 1 \text{ in.}$$

The maximum shearing stress will occur at the neutral axis which, for this symmetrical beam, is at its center. We will first determine $\bar{y}$ for the area above the center of the beam. Referring to Figure 8-9(b),

$$A_1 = 7.5 \text{ in.}^2, \quad y_1 = 1.25 \text{ in.},$$

$$A_2 = 4 \text{ in.}^2, \quad y_2 = 1 \text{ in.}$$

$$\bar{y} = \frac{7.5 \text{ in.}^2 (1.25 \text{ in.}) - 4 \text{ in.}^2 (1 \text{ in.})}{3.5 \text{ in.}^2}$$

$$= 1.54 \text{ in.}, \quad A' = 3.5 \text{ in.}^2$$

metric

$$V = \frac{4000 \text{ kg } (9.81 \text{ N/kg})}{2} = 19\,600 \text{ N}$$

$$I = \frac{7 \text{ cm } (12 \text{ cm})^3}{12} - \frac{5 \text{ cm } (10 \text{ cm})^3}{12}$$
$$= 591 \text{ cm}^4$$

$$t = 2 \text{ cm}$$

The maximum shearing stress will occur at the neutral axis which, for this symmetrical beam, is at its center. We will first determine $\bar{y}$ for the area above the center of the beam. Referring to Figure 8-8(b),

$$A_1 = 42 \text{ cm}^2, \quad y_1 = 3 \text{ cm},$$
$$A_2 = 25 \text{ cm}^2, \quad y_2 = 2.5 \text{ cm}$$

$$\bar{y} = \frac{42 \text{ cm}^2 (3 \text{ cm}) - 25 \text{ cm}^2 (2.5 \text{ cm})}{17 \text{ cm}^2}$$

$$= 3.74 \text{ cm}, \quad A' = 17 \text{ cm}^2$$

Now we can find the maximum shearing stress by use of Equation 8-4.

$$f_v = \frac{V(A'\bar{y})}{It}$$

$$= \frac{4000\,\text{lb}\,(3.5\,\text{in.}^2)(1.54\,\text{in.})}{20.6\,\text{in.}^4(1\,\text{in.})} = 1050\,\text{psi}$$

Now we can find the maximum shearing stress by using Equation 8-4.

$$f_v = \frac{V(A'\bar{y})}{It}$$

$$= \frac{19\,600\,N\left(\dfrac{17\,cm^2}{10\,000\,cm^2/m^2}\right)\left(\dfrac{3.74\,cm}{100\,cm/m}\right)}{\left(\dfrac{591\,cm^4}{10^8\,cm^4/m^4}\right)\left(\dfrac{2\,cm}{100\,cm/m}\right)}$$

$$f_v = 10.5\,MPa$$

In the case of I-shaped beams it is found that the flange contributes very little in resisting shearing forces as shown by the solid line of the shear diagram in Figure 8-9, which represents the shearing stress from the top to the bottom of the I beam. Therefore, for I-shaped beams (this includes wide-flange beams) the following approximation is generally used for computing the maximum shearing stress

$$f_v = \frac{V}{t_w d} \tag{8-5}$$

where f_v is the horizontal or vertical shearing stress, V is the vertical shearing force, t_w is the web thickness, and d is the depth of the beam. The approximate shear by Equation 8-5 is indicated by the dashed line in Figure 8-10.

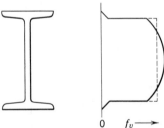

$$0 \qquad f_v \longrightarrow$$

FIGURE 8-10

Example

customary

Determine the shearing stress in a W8X15 wide-flange beam 30 ft long when

metric

Determine the shearing stress in a wide-flange beam 10 m long when

loaded with a 20 ton, 40,000 lb, concentrated load at its midpoint.

loaded with a 20 ton (20 t) load concentrated at its midpoint, if the beam's flange thickness is 0.622 cm and its depth is 20.6 cm.

Solution

customary

From Appendix 2 we find the web thickness of the beam t_w to be 0.245 in. and the depth of the beam d to be 8.12 in. Using Equation 8-5 we find

$$f_v = \frac{V}{t_w d} = \frac{20,000 \text{ lb}}{0.245 \text{ in.} (8.12 \text{ in.})} = 10,100 \text{ psi}$$

metric

The force of gravity acting on the 20 t load is

$$F = 20\,000 \text{ kg } (9.81 \text{ N/kg}) = 196 \text{ kN},$$
$$V = 196 \text{ kN}/2 = 98.1 \text{ kN}$$

Using Equations 8-5 we find

$$f_v = \frac{V}{t_w d} = \frac{98.1 \text{ kN}}{0.006\,22 \text{ m } (0.206 \text{ m})}$$
$$= 76.6 \text{ MPa}$$

8-3 STRESSES IN COMPOSITE BEAMS

A number of important materials do not have the same strengths in tension and compression. Consequently they do not function as efficiently as those that have the same strengths in tension and compression. Notable materials having this problem are cast iron, wood, and concrete. One way to overcome this problem while using such materials is to form a beam by combining two materials in a way that one of the materials helps to overcome the tensile or compressive weakness of the other. The most common example is the use of concrete and steel together in what is called reinforced concrete. Concrete is comparatively low in cost but its strength in tension is negligible. Obviously it would not be practical to use beams made of concrete alone because the tension at the bottom of the beams would cause them to fail at near zero load. However, if we imbed steel rods (*reinforcing rods*) in the beams near their bottoms, the reinforcing rods will absorb the tensile stresses and we will be able to take advantage of the economy of concrete. Such a beam, made of two materials which are inseparably joined together so that they bend as a unit, is called a *composite beam*.

When we combine two materials into a beam the neutral axis will be displaced from where it would be if the beam were made entirely from one or the other of the materials. One technique that may be used to determine the location of the neutral axis of a composite beam is to design an equivalent beam, made entirely of one or the other of the materials, that will react to loading in the same manner as the original composite beam. Then all of the calculations can be made in accordance with the previously developed beam theory.

Example

customary

For rigidity it is desired to use wooden beams in the construction of a house. However, to reduce the height of beam required it is decided to bond a steel plate to the bottom of the beam as shown in Figure 8-11. The maximum bending moment has been determined to be 10,000 in.-lb, b is 4 in., h_w is 6 in., h_s is 0.25 in. Determine the maximum compressive and tensile stresses in the composite beam.

metric

For rigidity it is desired to use wooden beams in the construction of a house. However, to reduce the height of beam required it is decided to bond a steel plate to the bottom of the beam as shown in Figure 8-11. The maximum bending moment has been determined to be 1500 N·m, b is 10 cm, h_w is 15 cm, h_s is 0.6 cm, E_{stl} is 210 GPa, and E_{wd} is 10.5 GPa. Determine the maximum compressive and tensile stresses in the wood and the maximum tensile stress in the steel.

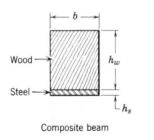

Composite beam

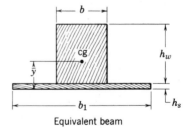

Equivalent beam

FIGURE 8-11

Solution

customary

Our first step is to design the equivalent beam by replacing the steel area by an area of wood that will act the same in bending as would the steel. We will call this area of wood the *transformed area*. We must not cause any change in the manner of bending of the original wood area. This means that the bend in the top of the beam must be the same as in the original composite beam. In the next chapter we will learn that the measure of the bend in the loaded beam, its curvature ρ, is related to the beam properties as follows

$$\frac{1}{\rho} = \frac{M}{EI}$$

The only value in the above equation that we may vary is I, which is directly proportional to b, h_w^3, and h_s^3. However, if we are to maintain the same curvature of the beam as required, we cannot change h_w or h_s. In the transformation it is only the b dimension that may be changed. Therefore, we will base our transformation on the equality

$$\frac{b_1}{b} = \frac{E_{stl}}{E_{wd}}$$

$$b_1 = b\frac{E_{stl}}{E_{wd}} \qquad (8\text{-}6)$$

where b_1 is the transformed width of the steel, b is the original width of the steel, E_{stl} and E_{wd} are the moduli of elasticity of the steel and wood, respectively.

Substituting the given values in Equa-

metric

Our first step is to design the equivalent beam by replacing the steel area by an area of wood that will act the same in bending as the steel. We will call this area of wood the **transformed area**. We must not cause any change in the manner of bending of the original wood area. This means that the bend in the top of the beam must be the same as in the original composite beam. In the next chapter we will learn that the measure of the bend in the loaded beam, its curvature ρ, is related to the beam properties as follows

$$\frac{1}{\rho} = \frac{M}{EI}$$

The only value in the above equation that may be varied is the moment of inertia I, which is directly proportional to b, h_w^3, and h_s^3. However, if we are to maintain the curvature of the beam as required, we cannot change the height dimensions. In the transformation it is only the b dimension that may be changed. Therefore, we will base our transformation on the equality

$$\frac{b_1}{b} = \frac{E_{stl}}{E_{wd}} \quad or \quad b_1 = b\frac{E_{stl}}{E_{wd}} \qquad (8\text{-}6)$$

where b_1 is the transformed width of the steel, b is the original width of the steel, E_{stl} and E_{wd} are the moduli of elasticity of the steel and wood, respectively.

Substituting the given values in

tion 8-6 we get

$$b_1 = b\frac{E_{stl}}{E_{wd}} = 4\ \text{in.}\ \frac{30,000,000\ \text{psi}}{1,500,000\ \text{psi}} = 80\ \text{in.}$$

We can now determine the properties of the equivalent beam. We will designate the original wood area 1 and the transformed steel area 2.

$$\bar{y} = \frac{A_1 y_1 + A_2 y_2}{A_1 + A_2}$$
$$= \frac{24\ \text{in.}^2(3.25\ \text{in.}) + 20\ \text{in.}^2(0.125\ \text{in.})}{44\ \text{in.}^2}$$
$$= 1.83\ \text{in.}$$

Now determine the moments of inertia about the centroid of the transformed area.

$$I_1 = \frac{bh_w^3}{12} + A_1 d_1^2 = \frac{4\ \text{in.}(6\ \text{in.})^3}{12}$$
$$+\ 24\ \text{in.}^2(3.25\ \text{in.} - 1.83\ \text{in.})^2 = 120\ \text{in.}^4$$

$$I_2 = \frac{b_1 h_s^3}{12} + A_2 d_2^2 = \frac{80\ \text{in.}(0.25\ \text{in.})^3}{12}$$
$$+\ 20\ \text{in.}^2(1.83\ \text{in.} - 0.125\ \text{in.})^2 = 58.2\ \text{in.}^4$$

$$I = I_1 + I_2 = 120^{114} + 58.2\ \text{in.}^4 = 178\ \text{in.}^4$$

Maximum compressive stress in the equivalent beam is

$$f_c = \frac{Mc}{I} = \frac{10,000\ \text{in.-lb}(6.25\ \text{in.} - 1.83\ \text{in.})}{178\ \text{in.}^4}$$
$$= 248\ \text{psi}$$

Maximum tensile stress in the equivalent beam is

$$f_t = \frac{Mc}{I} = \frac{10,000\ \text{lb-in.}(1.83\ \text{in.} - 0.25\ \text{in.})}{178\ \text{in.}^4}$$
$$= 88.8\ \text{psi}$$

If we were to calculate the stress in the bottom of the beam by the same method

Equation 8-6 we get

$$b_1 = b\frac{E_{stl}}{E_{wd}} = 10\ \text{cm}\ \frac{210\ GPa}{10.5\ GPa} = 200\ \text{cm}$$

We can now determine the properties of the equivalent beam. We will designate the original wood area 1 and the transformed steel area 2.

$$\bar{y} = \frac{A_1 y_1 + A_2 y_2}{A_1 + A_2}$$
$$= \frac{150\ cm^2(8.1\ cm) + 120\ cm^2(0.3\ cm)}{270\ cm}$$
$$= 4.63\ cm$$

Now determine the moments of inertia about the centroid of the transformed area.

$$I_1 = \frac{bh_w^3}{12} + A_1 d_1 = \frac{10\ cm(15\ cm)^3}{12}$$
$$+\ 150\ cm^2(8.1\ cm - 4.63\ cm)^2$$
$$= 4620\ cm^4$$

$$I_2 = \frac{b_1 h_s^3}{12} + A_2 d_2 = \frac{200\ cm(0.6\ cm)^3}{12}$$
$$+\ 120\ cm^2(4.63\ cm - 0.3\ cm)^2$$
$$= 2250\ cm^4$$

$$I = I_1 + I_2 = 4620\ cm^4 + 2250\ cm^4$$
$$= 6870\ cm^4$$

In meter units, $\bar{y} = 0.0463\ m$ and $I = 6.87 \times 10^{-5}\ m^4$. The maximum compressive stress in the wood is

$$f_c = \frac{Mc}{I} = \frac{1500\ N \cdot m(0.156\ m - 0.0463\ m)}{6.87 \times 10^{-5}\ m^4}$$
$$= 2.40\ MPa$$

The maximum tensile stress in the wood is

as the above stress calculations, our stress would be that distributed over the entire length b_1. To determine the actual stress in the steel we must multiply this distributed stress by the factor E_{stl}/E_{wd} so the equation for maximum stress in the steel becomes

$$f_{t_{stl}} = \frac{E_{stl}}{E_{wd}} \cdot f_t$$
$$= \frac{30{,}000{,}000 \text{ psi}}{1{,}500{,}000 \text{ psi}} \cdot 88.8 \text{ psi} = 1760 \text{ psi}$$

$$f_t = \frac{Mc}{I} = \frac{1500 \text{ N·m}(0.0463 \text{ m} - 0.006 \text{ m})}{6.87 \times 10^{-5} \text{ m}^4}$$
$$= 0.880 \text{ MPa}$$

If we were to calculate the stress in the bottom of the beam by the same method as the above stress calculations, our stress would be that distributed over the entire length b_1. To determine the actual stress in the steel we must multiply this distributed stress by the factor E_{stl}/E_{wd} so the equation for the stress in the steel becomes

$$f_{t_{stl}} = \frac{E_{stl}}{E_{wd}} \cdot f_t = \frac{210 \text{ GPa}}{10.5 \text{ GPa}} \cdot 0.880 \text{ MPa}$$
$$= 17.6 \text{ GPa}$$

8-4 ACCOMPLISHMENT CHECKLIST

From your study of this chapter you should now be able to

1. **Determine the normal bending stresses in statically determinate beams of uniform cross section.**

2. **Determine the section moduli of the above beams.**

3. **Use section moduli to select the most economical size and shape of beam for a given application.**

4. **Determine the vertical and horizontal shearing stresses in beams.**

5. **Determine the bending and shearing stresses in composite beams.**

8-5 SUMMARY

A condition of *pure bending* occurs at any section of a beam where there exists a bending moment but no shearing force. The plane on which there is no change of length when the beam is bent is the *neutral plane*. The *neutral axis* is the longitudinal axis of the neutral plane. The stress on any particle of a beam is proportional to its distance above or below the neutral axis. This is expressed by the equation

$$\frac{f_b}{c} = \frac{f_p}{y} \tag{8-1}$$

where f_b is the stress due to bending of the particle furthest above or below the neutral axis, c is the distance of the extreme fiber from the neutral axis, and f_p is the stress on any particle at a vertical distance y from the neutral axis. The bending stress f_b at any point in the beam is

$$f_b = \frac{My}{I} \tag{8-2}$$

where M is the bending moment of the beam at that point, y is the vertical distance from the neutral axis of the point, and I is the moment of inertia of the beam about the neutral axis. The maximum bending stress is given by

$$f_b = \frac{Mc}{I} = \frac{M}{S} \tag{8-3}$$

where S, $(= I/c)$, is the *elastic section modulus* of the beam.

The horizontal shearing stress equals the vertical shearing stress and is given by

$$f_v = \frac{VA'\bar{y}}{Ib} \tag{8-4}$$

where f_v is the shearing stress, V is the vertical shearing force, A' is the area of that portion of the cross section lying beyond the shear plane with respect to the neutral axis, $\bar{y}$ is the distance from the neutral axis to the centroid of A', I is the moment of inertia of the entire cross section of the beam, and b is the width of the beam at the shear plane. For I-shaped beams

$$f_v = \frac{V}{t_w d} \tag{8-5}$$

where t_w is the web thickness and d the depth of the beam.

The analysis of a composite beam is described in section 8-3.

PROBLEMS

In the problems that follow the weight of the beams is to be ignored unless otherwise specified.

bending stress problems

8-1C The figure shows a 1-15/16 diameter steel shaft mounted in two ball bearings which permit it to move endways and to bend. If L is 48 in., a is 12 in., and P is 2000 lb, what is the maximum bending stress developed in the beam? E for steel is 30,000,000 psi.

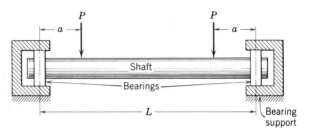

PROBLEM 8-1

8-1M *The figure shows a 50 mm steel shaft mounted in two ball bearings which permit it to move endways and to bend. If L is 1.2 m, a is 30 cm, and P is 800 kg, what is the maximum bending stress developed in the beam? E for steel is 207 GPa.*

8-2C Determine the maximum load P the shaft in Problem 8-1C could support if the maximum allowable bending stress were 20,000 psi.

8-2M *Determine the maximum load P the shaft in Problem 8-1M could support if the maximum allowable bending stress were 140 MPa.*

8-3C Determine the maximum bending stresses in the wood floor joist shown if it supports a uniform load of 75 lb/ft with $L = 18$ ft, $b = 4$ in., and $h = 12$ in. E for the wood is 1,500,000 psi.

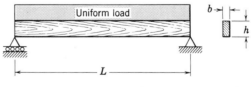

PROBLEM 8-3

8-3M *Determine the maximum bending stresses in the wood floor joist shown if it supports a uniform load of 50 kg/m with L = 6 m, b = 100 mm, and h = 300 mm. E for the wood is 10 GPa.*

8-4C Determine the maximum bending stress in the beam of Problem 8-3C if, in addition to the uniform load, there were a concentrated load of 1000 lb 4 ft from its left end.

8-4M *Determine the maximum bending stress in the beam of Problem 8-3M if, in addition to the uniform load, there were a concentrated load of 500 kg 1.2 m from its left end.*

8-5C Determine the maximum permissible uniform load for the beam in Problem 8-3C if the allowable tensile stress is 1100 psi.

8-5M *Determine the maximum permissible uniform load for the beam in Problem 8-3M if the allowable bending stress in tension is 7.6 MPa.*

8-6C A structural glued laminated timber beam supported only at its ends is 100 ft long. It is 10-3/4 in. wide by 60 in. deep and is designed to carry a maximum load of 960 lb/ft including its own weight. It is desired to suspend a scoreboard from its midpoint. Determine the maximum permissible weight for the scoreboard if the only uniform load it will carry is its own weight of 157 lb/ft.

8-6M *A structural glued laminated timber beam supported only at its ends is 30 m long. It is 27.3 cm wide by 152 cm deep and is designed to carry a maximum load of 1430 kg/m including its own weight. It is desired to suspend a scoreboard from its midpoint. Determine the maximum permissible weight for the scoreboard if the only uniform load it will carry is its own weight of 234 kg/m.*

8-7C The air conditioning unit shown in the figure weighs 680 lb. It is to be supported on the two equal-legged structural angles as shown. L is 24 ft, a is 6 ft, and F_b = 20,000 psi. Determine the most economical angle for this purpose from those in Appendix 2.

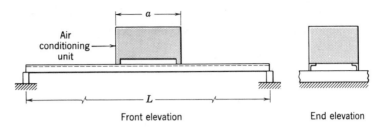

Front elevation   End elevation

PROBLEM 8-7

8-7M *The air conditioning unit shown in the figure weighs 320 kg. It is to be supported by two equal-legged structural angles as shown. L is 8 m, a is 2 m, and $F_b = 138$ MPa. Determine the most economical angle for this purpose from those in Appendix 2.*

8-8C The figure shows a wooden beam of a type frequently used to assist in hauling heavy or bulky items up to lofts or upper floors of buildings. In use a pulley would be attached to the hook and the load would be lifted by a rope or cable passed through the pulley. The load, represented by P, is to be limited to 1500 lb. Determine the dimensions of the beam required if $h = 2b$ and the allowable bending stress is 960 psi. L is 5 ft and a is 0.5 ft.

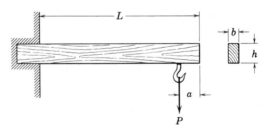

PROBLEM 8-8

8-8M *The figure shows a wooden beam of a type frequently used to assist in hauling heavy or bulky items up to lofts or upper floors of buildings. In use a pulley would be attached to the hook and the load would be lifted by a rope or cable passed through the pulley. The load, represented by P, is to be limited to 750 kg. Determine the dimensions of the beam required if $h = 2b$ and the allowable bending stress is 6.6 MPa, L is 1.5 m, and a is 15 cm.*

8-9C   Determine the maximum stress in section A—A of the cast-iron ($E = 17 \times 10^6$ psi) C-clamp when P is 200 lb, w is 3/8 in., h is 1/2 in., t is 5/64 in., and b is 3/4 in.

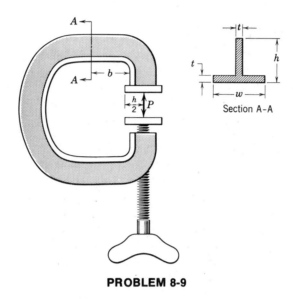

PROBLEM 8-9

8-9M   *Determine the maximum stress in section A—A of the cast-iron ($E = 117\,GPa$) C-clamp when P is 1 kN, w is 9.5 mm, h is 12.5 mm, t is 2 mm, and b is 20 mm.*

8-10C   The figure shows a lever used in a control mechanism. It is made of 5456-H343 aluminum which has an allowable bending stress of 39 ksi. L_1 is 1 in., L_2 is 3 in., h is 3/8 in., b is 5/64 in., and the hole diameter d is 3/16 in. Determine the maximum permissible values of P_1 and P_2.

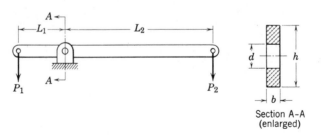

PROBLEM 8-10

8-10M *The figure shows a lever used in a control mechanism. It is made of 5456-H343 aluminum which has an allowable bending stress of 269 MPa. L_1 is 3 cm, L_2 is 8 m, h is 1 cm, b is 2 mm, and the hole diameter d is 0.5 cm. Determine the maximum permissible values of P_1 and P_2.*

8-11C The inverted T-beam in the figure is loaded with a variable load that increases uniformly from zero at the left end to $w = 150$ lb/ft at the right end. L is 8 ft, b is 5 in., h is 5 in., and t is 1/2 in. Determine the maximum bending stress.

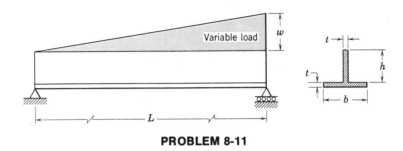

PROBLEM 8-11

8-11M *The inverted T-beam in the figure is loaded with a variable load that increases uniformly from zero at the left end to $w = 200$ kg/m at the right end. L is 3 m, b is 12 cm, h is 12 cm, and t is 1 cm. Determine the maximum bending stress.*

8-12C At what uniform load on the W12X65 wide-flange beam would it be stressed to its allowable bending stress of 20,000 psi if L is 40 ft and E is 29,000,000 psi.

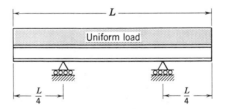

PROBLEM 8-12

8-12M *At what uniform load on the $300 \times 150 - 32$ wide-flange beam would it be stressed to its allowable bending stress of 140 MPa if L is 12 m and E is 200 GPa.*

211

8-13C The beam in the figure is an aluminum square tube, 3 in. × 3 in. with a wall thickness of 0.219 in. The beam is under a variable load and a concentrated load as shown; w is 30 lb/ft, P is 250 lb, L_1 is 20 ft, and L_2 is 8 ft. Determine the maximum bending stress in the beam.

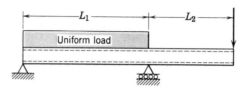

PROBLEM 8-13

8-13M *The beam in the figure is an aluminum square tube, 75 mm × 75 mm with a wall thickness of 6 mm. The beam is under a variable load and a concentrated load as shown. L_1 is 6 m, L_2 is 2.5 m, w is 45 kg/m, and P is 100 kg. Determine the maximum bending stress in the beam.*

shearing stress problems

8-14C Determine the maximum shearing stress in the beam of Problem 8-3C.

8-14M *Determine the maximum shearing stress in the beam of Problem 8-3M.*

8-15C Determine the maximum shearing stress in the beam of Problem 8-4C.

8-15M *Determine the maximum shearing stress in the beam of Problem 8-4M.*

8-16C Determine the maximum permissible concentrated load 6 ft from the left end of the beam in Problem 8-3C if the allowable shear load is 70 psi. Compare your answer with the answer for Problem 8-3C.

8-16M *Determine the maximum permissible concentrated load 2 m from the left end of the beam in Problem 8-3M if the allowable shear load is 483 kPa. Compare your answer with the answer for Problem 8-3M.*

8-17C A simple span beam made of structural glued laminated timber is 5-1/8 in. wide by 21 in. high and is 20 ft long. The allowable design stresses are $F_b = 2400$ psi and $F_v = 165$ psi. $E = 1,800,000$ psi. Determine the maximum concentrated load that may be carried at its midpoint if the beam itself weighs 19.62 lb/ft.

8-17M *A simple span beam made of structural glued laminated timber is 13 cm wide by 53.3 cm high and is 6 m long. The allowable design stresses are $F_b = 16.5$ MPa and $F_v = 1.14$ MPa. $E = 12.4$ GPa. Determine the maximum concentrated load that may be carried at its midpoint if the beam itself weighs 29.2 kg/m.*

8-18C Determine if the dimensions determined for the beam in Problem 8-8C are satisfactory if the allowable shearing stress for the beam is 75 psi.

8-18M *Determine if the dimensions determined for the beam in Problem 8-8M are satisfactory if the allowable shearing stress for the beam is 520 kPa.*

8-19C Determine if the C-clamp of Problem 8-9C is overdesigned, underdesigned, or just right for a maximum 200 lb load P, if the material is cast iron with $E = 14,500,000$ psi, $F_t = 20,000$ psi, $F_c = 74,000$ psi, and $F_v = 28,000$ psi.

8-19M *Determine if the C-clamp of Problem 8-9M is overdesigned, underdesigned, or just right if the material is cast iron with $E = 100$ GPa, $F_t = 140$ MPa, $F_c = 510$ MPa, and $F_v = 190$ MPa.*

8-20C Determine if a wood beam 5-1/8 in. wide by 10-1/2 in. high being used as a simple beam with a 30 ft span to carry a 400 lb/ft load (including the beam weight) is safe if the allowable shear stress F_v is 165 psi.

8-20M *Determine if a wood beam 13 cm wide by 26.7 cm high being used as a simple beam with a 9.5 m span to carry a 500 kg/m load (including beam weight) is safe if the allowable shear stress F_v is 1.15 MPa.*

8-21C Determine the shearing stress in a W8X15 wide-flange beam 30 ft long when loaded with a uniform load of 250 lb/ft. Use both the exact and approximate equations.

8-21M   *Determine the shearing stress in a $200 \times 100 - 21.3$ wide-flange beam with a 9.5 m span when used as a simple beam carrying a uniform load of 350 kg/m. Use both the exact and approximate equations.*

8-22C   Three steel plates 6 in. $\times$ 1/2 in. are welded together as in the figure to make a simple beam 30 ft long. It is loaded with a uniform load of 200 lb/ft over its entire length and two concentrated loads of 2000 lb each applied at points 10 ft from either end of the beam. Determine the maximum shearing stress by the approximate method.

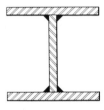

PROBLEM 8-22

8-22M   *Three steel plates 15 cm $\times$ 1.25 cm are welded together as in the figure to make a beam 10 m long. It is simply mounted and loaded with a uniform load of 450 kg/m over its entire length and two concentrated loads of 1 000 kg each at points 3 m from either end of the beam. Determine the maximum shearing stress by the approximate method.*

8-23C   If a steel beam 80 ft long made of ASTM-A36 steel were used to support, as a simple beam, a uniform load of 14,000 lb/ft, what would be the required web thickness to be safe in shear if the depth of the beam was 30 in? Use the approximate method.

8-23M   *If a steel beam 25 m long were used to support, as a simple beam, a uniform load of 21 000 kg/m, what would be the required web thickness to be safe in shear if the depth of the beam was 800 mm? Use the approximate method.*

composite beam problems

8-24C   A beam is formed by sandwiching a 1 in. $\times$ 6 in. aluminum plate having an E of 10,000,000 psi between two 1 in. $\times$ 6 in. white pine boards having an E of 1,000,000 psi, as shown in the figure. The members of the beam are inseparably bonded. Determine the maximum bending stress in the pine members of the beam when it is used as a simple beam 10 ft long and carries a uniform load of 50 lb/ft.

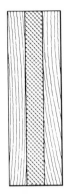

PROBLEM 8-24

8-24M *A beam is formed by bonding a 25 mm × 300 mm aluminum plate having an E of 69 GPa between two 25 mm × 300 mm white pine boards having an E of 6.9 GPa as shown in the figure. Determine the maximum bending stress in the pine members of the beam when it is used as a simple beam 3 m long and carries a uniform load of 75 kg/m.*

8-25C Determine the stress in the aluminum in Problem 8-24C.

8-25M *Determine the stress in the aluminum in Problem 8-24M.*

8-26C Same as Problem 8-24C except that the center piece is wood and the outside pieces are aluminum. Transform the wood.

8-26M *Same as Problem 8-24M except that the center piece is wood and the outside pieces are aluminum. Transform the wood.*

8-27C Three 1 in. × 8 in. boards are bonded together as shown. The two outside boards are white pine with $F_b = 1000$ psi, $F_v = 80$ psi, and $E = 1,000,000$ psi. The middle board is Douglas fir with $F_b = 1450$ psi, $F_v = 80$ psi, and $E = 1,600,000$ psi. Determine the maximum uniform load the beam can safely support as a simple beam 20 ft long.

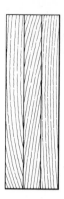

PROBLEM 8-27

8-27M   *Three 25 mm × 200 mm boards are bonded together as shown. The two outside boards are white pine with F_b = 6.9 MPa, F_v = 550 kPa, and E = 6.9 GPa. The middle board is Douglas fir with F_b = 10 MPa, F_v = 550 kPa, and E = 11 GPa. Determine the maximum uniform load the beam can safely support as a simple beam 6 m long.*

8-28C   Solve Problem 8-26C with the beams longer cross-section dimension horizontal.

8-28M   *Solve Problem 8-26M with the beams longer cross-section dimension horizontal.*

8-29C   A composite beam is formed by inseparably bonding a wooden beam 3 in. wide by 5 in. high to a 1/2 in. by 3 in. steel bar as shown. The beam is a cantilever which extends 6 ft beyond its support. Determine the largest concentrated load that can be safely placed on the beam's free end. The wood is Douglas fir with F_b = 1450 psi, F_v = 80 psi, and E = 1,600,000 psi. The steel is SAE 1025 with F_b = 40,000 psi, F_v = 30,000 psi, and E = 29,000,000 psi.

PROBLEM 8-29

8-29M *A composite beam is formed by inseparably bonding a wooden beam 75 mm wide by 125 mm high to a 75 mm by 15 mm steel bar as shown. The beam is a cantilever which extends 2 m beyond its support. Determine the largest concentrated load that can be safely placed on the beam's free end. The wood is Douglas fir with $F_b = 10$ MPa, $F_v = 550$ kPa, and $E = 11$ GPa. The steel is carbon steel with $F_b = 275$ MPa, $F_v = 210$ MPa, and $E = 200$ GPa.*

8-30C A 6 in. × 6 in. composite beam is formed as shown with the steel having a uniform thickness of 3/8 in. The specifications for the wood and steel are the same as in Problem 8-29C. Determine the maximum uniform load that can safely be placed on the beam if it is used as a cantilever 12 ft long.

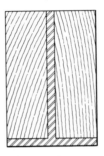

PROBLEM 8-30

8-30M *A 15 cm × 15 cm composite beam is formed as shown with the steel having a uniform thickness of 1 cm. The specifications for the wood and steel are the same as in Problem 8-29M. Determine the maximum uniform load that can be safely placed on the beam if it is used as a cantilever 3.5 m long.*

CHAPTER
OBJECTIVES

The object of this chapter is to learn

1. The terms used in describing the deflection of beams.

2. The theorems governing the moment-area method of determining beam deflections.

3. The procedure using the moment-area theorems to determine beam deflections.

4. How to apply the procedure to beams of different types.

5. How to apply information developed by the moment-area method in determining deflections of beams by superposition.

CHAPTER —————————————— 9
DEFLECTION OF BEAMS

The last chapter was devoted to the consideration of the strength of beams. The purpose was to learn how to select beams that would take the required loads without failing. Beams can be stressed beyond their elastic limits into the plastic range and still not fail. The determination of the point of plastic failure is the subject of a more advanced course.

In many situations beams which do not fail either elastically or plastically may still be unsatisfactory because they deflect too much under the loads they must support. Too much deflection of bridge spans can cause spalling of the road surface. Too much deflection of beams in buildings can cause cracking of plaster in ceilings or the formation of ponds of rain water on roofs. Shafts in machines can reach critical speeds where they will suddenly fail due to destructive vibrations caused by excessive deflection. The beam portions of machines producing parts to accuracies in the tenths of thousands of inches or in micrometers must be extremely rigid. Excessive deflection would cause failure of the machines to do the job for which they were designed.

It is very important to know how to determine the amount of deflection in beams due to different types of loading.

9-1 BEAM DEFLECTION

When a beam has no load on it, even its own weight, it is assumed to be straight and its neutral axis is a straight line as in Figure 9-1(a). When a load is placed on it, it bends as is shown with great exaggeration in part (b) of the

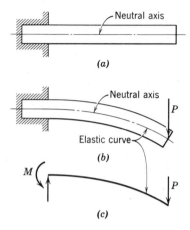

FIGURE 9-1

figure. The shape of the neutral axis (its curve) is called the *elastic curve* of the loaded beam. When we draw the elastic curves of beams we will draw them as single heavy lines as in Figure 9-1(c).

There are a number of methods of determining beam deflections. Of these, two have been selected for presentation here: (1) The moment-area method and (2) the superposition method, which is an application of the information developed by other methods such as the moment-area method. With the ability to apply these methods you will be able to handle the majority of straightforward problems.

9-2 MOMENT-AREA THEOREMS

The moment-area method is based on the curvature of the elastic curve of a beam which is exposed to a bending moment M. The curvature is

$$\frac{1}{\rho} = \frac{M}{EI} \tag{9-1}$$

where ρ is the *radius of curvature* of any infinitely small portion of the elastic curve, $1/\rho$ is the *curvature* of that portion of the elastic curve, E is the modulus of elasticity of the beam material, and I is the moment of inertia of the cross section of the beam. The product EI is commonly referred to as the *flexural rigidity* of the beam. Using this relationship an ingenious German engineer, Otto Mohr, developed the moment-area method. The essence of his method is contained in two theorems.

The *first moment-area theorem* may be stated as follows:

The angle (in radians) between the tangents to the elastic curve at any two points A and B is equal to the area of the moment diagram of the beam between the ordinates of those same two points, divided by the flexural rigidity EI of the beam.

This theorem is useful in determining the slope of the elastic curve at any point. Every elastic curve has at least one point at which its tangent is horizontal. Take this as one of your points and the point for which you desire

Table 9.1

	Shape	Area	Centroid Location
Rectangular		bh	$\bar{x} = \dfrac{b}{2}, \quad \bar{y} = \dfrac{h}{2}$
Triangular		$\dfrac{1}{2}bh$	$\bar{x} = \dfrac{b}{3}, \quad \bar{y} = \dfrac{h}{3}$
Parabolic		$\dfrac{1}{3}bh$	$\bar{x} = \dfrac{b}{4}, \quad \bar{y} = \dfrac{3h}{10}$
Parabolic		$\dfrac{2}{3}bh$	$\bar{x} = \dfrac{3b}{8}, \quad \bar{y} = \dfrac{2h}{5}$

the slope as the other, proceed according to the theorem, and the angle you determine will be the desired slope.

In applying the moment-area theorems we must know how to determine the areas and locate the centroids of the basic shapes of which the areas of the moment diagrams are composed. These are given in Table 9-1.

Example

customary

Determine the slope at the end of a cantilever beam which is 15 ft long, is made of steel with $E = 29{,}000{,}000$ psi, has a moment of inertia I of 30.8 in.^{4}, and carries a load P of 3000 lb at its free end.

metric

Determine the slope at the end of a cantilever beam which is 5 m long, is made of steel with $E = 200$ GPa, has a moment of inertia I of $1.28 \times 10^{-5} m^4$, and carries a load P of 1500 kg.

Solution

customary

For the cantilever beam specified the elastic curve may be drawn as shown in Figure 9-2. Note that the tangent to the

metric

For the cantilever beam specified the elastic curve may be drawn as shown in Figure 9-2. Note that the tangent to

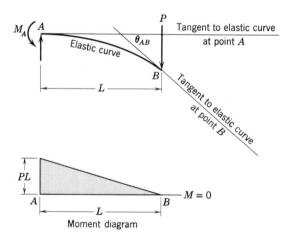

FIGURE 9-2

elastic curve at point A is horizontal. Therefore, the angle θ_{AB} formed by the tangent at point B at the end of the beam and the horizontal point A tangent reveals the angle of the slope of the end of the beam.

The moment-area theorem tells us that θ_{AB} is equal to the area of the moment diagram between points A and B divided by EI. From the moment diagram in the figure

$$\theta_{AB} = \frac{\text{Area}}{EI} = \frac{PL^2/2}{EI} = \frac{1}{2}(3000 \text{ lb})(180 \text{ in.})^2$$
$$\times \; 129{,}000{,}000 \text{ psi } (30.8 \text{ in.}^4)$$
$$= 5.44 \times 10^{-2} \text{ rad} \quad \text{or} \quad 3.12°$$

the elastic curve at point A is horizontal as it must be because at this point the beam emerges horizontally from its support. Therefore, the angle θ_{AB} formed by the tangent to the elastic curve at point B, at the end of the beam, and the horizontal point A tangent reveals the angle of the slope of the end of the beam.

The moment-area theorem tells us that θ_{AB} is equal to the area of the moment diagram between points A and B divided by EI. Using the moment diagram in the figure

$$\theta_{AB} = \frac{\text{Area}_{AB}}{EI} = \frac{PL^2/2}{EI}$$
$$= \frac{(1500 \text{ kg} \times 9.81)(5 \text{ m})^2/2}{200 \text{ GPa } (1.28 \times 10^{-5} \text{ m}^4)} = 7.19$$
$$\times \; 10^{-2} \text{ rad} \quad \text{or} \quad 4.12°$$

where Area_{AB} is the area of the moment diagram between points A and B.

Note that the elastic curves we draw are simply our visualization of the curves the beams take under load with the curves and deflections greatly exaggerated as an aid in solving the problems. The actual angles of deflection of beams loaded within the elastic range will be quite small as in the foregoing examples.

The second *moment-area theorem* may be stated

If a tangent is drawn to an elastic curve at a point A, the vertical distance of any other point B on the elastic curve from that tangent is the moment with respect to B of the area of the moment diagram between the ordinates of points A and B divided by the flexural rigidity EI.

If, in the application of the second moment-area theorem, we select as point A a point on the elastic curve that has a horizontal tangent, the application of the theorem brings us directly to the vertical distance between A and any other point B on the curve.

Example

customary

Determine the vertical deflection of the end of a 15 ft cantilever beam made of steel with an E of 29,000,000 psi, having an I of 30.8 in.^{4}, and carrying a load of 3000 lb at its free end.

metric

Determine the vertical deflection of the end of a 5 m cantilever beam made of steel with an E of 200 GPa, having an I of $1.28 \times 10^{-5} m^4$ which carries a load of 1500 kg at its free end.

Solution

customary

A tangent to the elastic curve at A in Figure 9-3 is horizontal. We can find the vertical deflection at the free end of the beam by

$$\Delta_B = d_{B/A} = \frac{\text{Area}_{AB}(\bar{x})}{EI}$$
$$= \frac{PL^2/2(2L/3)}{EI} = \frac{PL^3}{3EI}$$

where Δ_B is the deflection of point B, $d_{B/A}$ is the vertical distance between the

metric

A tangent to the elastic curve at A in Figure 9-3 is horizontal. We can find the vertical deflection at the free end of the beam by

$$\Delta_B = d_{B/A} = \frac{\text{Area}_{AB}(\bar{x})}{EI}$$
$$= \frac{PL^2/2(2L/3)}{EI} = \frac{PL^3}{3EI}$$

where Δ_B is the deflection of point B, $d_{B/A}$ is the vertical distance between the

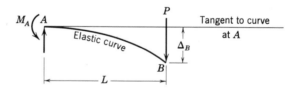

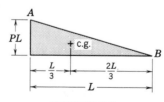

FIGURE 9-3

deflected point B and the tangent to point A of the elastic curve, and Area$_{AB}$ is the area of the moment diagram between the ordinates passing through points A and B

$$\Delta_B = \frac{3000 \text{ lb } (15 \text{ ft} \times 12 \text{ in./ft})^3}{3(29,000,000 \text{ psi})(30.8 \text{ in.}^4)} = 6.53 \text{ in.}$$

deflected point B and the tangent to point A of the elastic curve, and Area$_{AB}$ is the area of the moment diagram between the ordinates passing through points A and B.

$$\Delta_B = \frac{1500 \text{ kg} \times 9.81 \, (4 \text{ m})^3}{3\,(200 \text{ GPa})(1.28 \times 10^{-5} \text{ m}^3)}$$
$$= 0.123 \text{ m} \quad or \quad 12.3 \text{ cm}$$

9-3 MOMENT-AREA APPLICATIONS

The examples of the previous section were of the simplest possible systems. The purpose there was to clarify the statements of the theorems. We will now examine some more difficult systems and develop some techniques for handling them.

In the previous discussion it is suggested that we start by selecting a point on our elastic curve which has a horizontal tangent. For cantilever beams and simply supported beams with symmetrical loading such points are easy to locate by examining the elastic curve. However, in other cases, such as that shown in Figure 9-4, it is not that easy. The procedure is as follows.

STEP 1.

Draw the free-body diagram and determine the reactions at the supports.

STEP 2.

Draw the elastic curve. This should be smooth, continuous, and with no abrupt changes. If the shape of the curve is not known, it may be assumed. Label the angles and distances as shown. The lowest point M must be estimated.

STEP 3.

Draw the moment diagram as shown. The height of the moment-area diagram at M is h. The values of moments at important points should be shown.

DEFLECTION OF BEAMS

STEP 4.

Determine the distance $d_{B/A}$, from point B to the tangent drawn to the elastic curve at A by the second moment-area theorem.

$$\text{Area}_{AB} = \frac{1}{2}bh = \frac{1}{2}L\left(Pa - \frac{Pa^2}{L}\right) = \frac{Pa}{2}(L-a)$$

$$\bar{x}_B = \frac{L+a}{3}$$

$$d_{B/A} = \frac{\text{Area}_{AB}\bar{x}_B}{EI} = \frac{Pa/2(L-a)((L+a)/3)}{EI} = \frac{Pa(L^2-a^2)}{6EI}$$

STEP 5.

Determine θ_{AM} from the geometry of the elastic curve. For very small angles such as are encountered in beam deflections $\tan \theta_A$ will be very close in value to θ_A. Since $\theta_A = \theta_{AM}$ (similar triangles)

$$\theta_{AM} = \frac{d_{B/A}}{L} = \frac{Pa}{6EI}\left(\frac{L^2-a^2}{L}\right)$$

STEP 6.

We can also determine θ_{AM} by the first moment-area theorem as follows.

$$\theta_{AM} = \frac{\text{Area}_{AM}}{EI}$$

where Area_{AM} is the area of the moment diagram between points A and M. The height of Area $_{AM}$ may be found by proportional triangle geometry

$$\frac{h}{x} = \frac{Pa - Pa^2/L}{(L-a)} = \frac{Pa}{L} \qquad \text{thus} \quad h = \frac{Pax}{L}$$

We can now determine the area of the moment diagram between A and M.

$$\text{Area}_{AM} = \frac{1}{2}xh = \frac{1}{2}x\left(\frac{Pax}{L}\right) = \frac{Pax^2}{2L}$$

then

$$\theta_{AM} = \frac{\text{Area}_{AM}}{EI} = \frac{Pax^2}{2LEI}$$

STEP 7.

Equate the results of Steps 5 and 6 and solve for x, the distance from point A to the point of maximum deflection.

$$\theta_{AM} = \theta_{AM}$$

$$\frac{Pa}{6\,EI}\left(\frac{L^2 - a^2}{L}\right) = \frac{3\,Pax^2}{6\,LEI}$$

$$L^2 - a^2 = 3x^2$$

$$x^2 = \frac{L^2 - a^2}{3}$$

$$x = \sqrt{\frac{L^2 - a^2}{3}}$$

which is the horizontal distance between points A and M.

STEP 8.

Determine the maximum deflection of the beam. We can do this directly by determining the distance $d_{M/A}$ of point A from the horizontal tangent through M.

$$d_{M/A} = \text{Area}_{AM}\bar{x}_A / EI$$

$$\Delta_M = d = \frac{Pax^2}{2\,LEI}\left(\frac{2x}{3}\right) = \frac{Pax^3}{3\,LEI}$$

We now substitute the value of x at M in the above equation.

$$\Delta_M = \frac{Pa((L^2 - a^2)/3)^{3/2}}{3\,LEI} = \frac{Pa(L^2 - a^2)^{3/2}}{9\sqrt{3}\,LEI}$$

Let us apply this equation to the determination of the deflection of a beam having a single concentrated load not at its center.

Example

customary

Determine the maximum deflection of a beam such as shown in Figure 9-4. The beam is 21 ft long with a concentrated

metric

Determine the maximum deflection of an aluminum beam such as is shown in Figure 9-4. The beam is 6 m long with a

DEFLECTION OF BEAMS

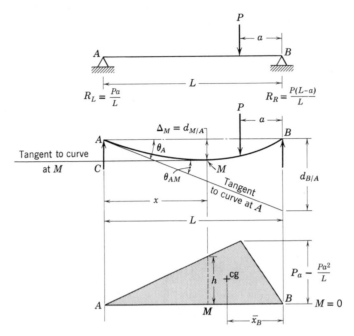

FIGURE 9-4

load of 2000 lb 8 ft from its right-hand end. It is made of aluminum having an E of 10,000,000 psi and it has an I of 109 in.^{4}

concentrated load of 1 000 kg at 2 m from its right-hand end. E for the beam material is 69 GPa and I for the cross section of the beam is $4.54 \times 10^{-5}\,m^4$.

Solution

customary

$$\Delta_M = \frac{Pa\,(L^2 - a^2)^{3/2}}{9\sqrt{3}\,LEI}$$

In substituting the given dimensions in the above equation we must be consistant. We therefore will use inches throughout.

metric

$$\Delta_M = \frac{Pa\,(L^2 - a^2)^{3/2}}{9\sqrt{3}\,LEI}$$

$$\Delta_M = \frac{(1\,000\,kg \times 9.81\,kg/N)(2\,m)((6\,m)^2 - (2\,m)^2)^{3/2}}{9\sqrt{3}(6\,m)(69\,GPa)(4.54 \times 10^{-5}\,m^4)}$$

$$\Delta_M = 0.0121\,m \quad or \quad 12.1\,mm$$

$$L = 21 \text{ ft } (12 \text{ in./ft}) = 252 \text{ in.,}$$
$$a = 8 \text{ ft } (12 \text{ in./ft}) = 96 \text{ in.}$$

$$\Delta_M = \frac{2000 \text{ lb } (96 \text{ in.})((252 \text{ in.})^2 - (96 \text{ in.})^2)^{3/2}}{9\sqrt{3}(252 \text{ in.})(10,000,000 \text{ psi})(109 \text{ in.}^4)}$$

$$= 0.567 \text{ in.}$$

Thus far we have considered beams under single concentrated loads only. To round out our study we will now examine a beam that has a combination of a uniform load and a concentrated load. We will use the foregoing step-by-step procedure to determine its maximum deflection.

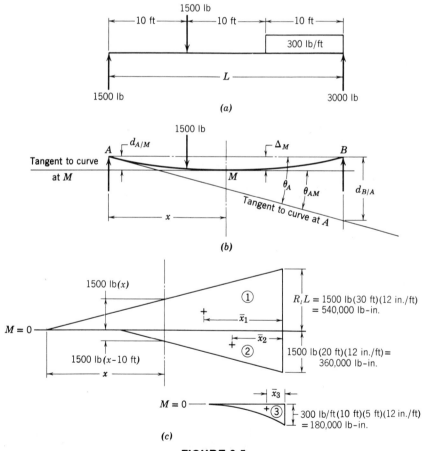

FIGURE 9-5

Example

customary

Determine the maximum deflection of the beam of which the free-body diagram is shown in Figure 9-5(a). The beam is a W10X33 steel beam having an E of 29,000,000 psi and an I of 171 in.[4]

metric

Determine the maximum deflection of the beam of which the free-body diagram is shown in Figure 9-6(a). The beam is a steel beam having an E of 200 GPa and an I of $7.12 \times 10^{-5}\,m^4$.

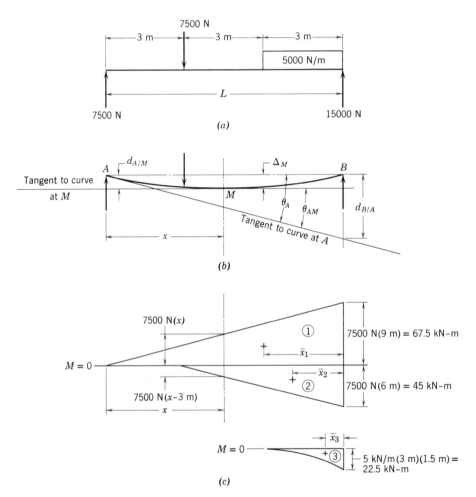

(a)

(b)

(c)

FIGURE 9-6

Solution

customary   **metric**

STEP 1.

Shown in Figure 9-5(a).

STEP 1.

Shown in Figure 9-6(a).

STEP 2.

Draw the elastic curve. Here we can only estimate where the lowest point M will be. It appears that it will be between the uniform load and the concentrated load. Label the angles and distances as shown in Figure 9-5(b).

STEP 2.

Draw the elastic curve. Here we can only estimate where the lowest point M will be. It appears that it will be between the uniform load and the concentrated load. Label the angles and distances as shown in Figure 9-6(b).

STEP 3.

Draw the moment diagram. Up to this point we have been drawing moment diagrams as a single curve. It is very difficult to calculate the areas and centroids of such diagrams for beams having combined loading. Moment diagrams may be drawn by parts as shown in Figure 9-5(c), where the moment of each force and load is shown separately. The large triangle above the $M = 0$ line represents the bending moment caused by the left-hand reaction, while the smaller triangle below the line represents the bending moment of the 1500 lb load which causes a negative moment. To draw the bending moment diagram for the uniform load without overlapping one of the previously drawn diagrams we establish another $M = 0$ line and draw the diagram as shown. Thus, the total moment-area to the left of any point on the beam is the

STEP 3.

Draw the moment diagram. Up to this point we have been drawing moment diagrams as a single curve. It is very difficult to calculate the areas and centroids of such diagrams for beams that have combined loading. Moment diagrams may be drawn by parts as shown in Figure 9-6(c) where the moment of each force and load is shown separately. The large triangle above the M = 0 line represents the bending moment caused by the left-hand reaction, while the smaller triangle below the line represents the bending moment of the 7500 N load which causes a negative moment. To draw the bending moment diagram for the uniform load without overlapping one of the previously drawn diagrams we establish another M = 0 line and draw the diagram as shown. Thus the

DEFLECTION OF BEAMS

sum of the areas above the line minus the sum of those below the line to the left of the point.

STEP 4.

Determine the distance $d_{B/A}$ by the second moment-area theorem. Here we handle the areas individually as follows. Note that the moments of those areas above the line are positive while those below the line are negative.

$$\text{Area}_1 = \tfrac{1}{2}(360 \text{ in.})(540{,}000 \text{ lb-in.})$$
$$= 97.2 \times 10^6 \text{ lb-in.}$$

$$\text{Area}_2 = \tfrac{1}{2}(240 \text{ in.})(360{,}000 \text{ lb-in.})$$
$$= 43.2 \times 10^6 \text{ lb-in.}$$

$$\text{Area}_3 = \tfrac{1}{2}(120 \text{ in.})(180{,}000 \text{ lb-in.})$$
$$= 7.2 \times 10^6 \text{ lb-in.}$$

$$\bar{x}_1 = \frac{360 \text{ in.}}{3} = 120 \text{ in.}, \quad \bar{x}_2 = \frac{240 \text{ in.}}{3} = 80 \text{ in.},$$

$$\bar{x}_3 = \frac{120 \text{ in.}}{4} = 30 \text{ in.}$$

$$d_{B/A} = \frac{\text{Area}_1 \bar{x}_1 - \text{Area}_2 \bar{x}_2 - \text{Area}_3 \bar{x}_3}{EI}$$

$$= \frac{(97.2 \times 10^6)(120 \text{ in.}) - (43.2 \times 10^6)(80 \text{ in.})}{(29 \times 10^6)(171 \text{ in.}^4)}$$

$$- \frac{(7.2 \times 10^6)(30 \text{ in.})}{(29 \times 10^6)(171 \text{ in.}^4)}$$

$$= 1.61 \text{ in.}$$

total moment-area to the left of any point on the beam is the sum of the areas above the line minus the sum of those below the line to the left of the point.

STEP 4.

Determine the distance $d_{B/A}$ by the second moment-area theorem. Here we handle the areas individually as follows. Note that the moments of those areas above the line are positive while those below the line are negative.

$$\text{Area}_1 = \tfrac{1}{2}(9 \text{ m})(67.5 \text{ kN} \cdot m)$$
$$= 304 \text{ kN} \cdot m^2,$$
$$\bar{x}_1 = \frac{9 \text{ m}}{3} = 3 \text{ m}$$

$$\text{Area}_2 = \tfrac{1}{2}(6 \text{ m})(45 \text{ kN} \cdot m)$$
$$= 135 \text{ kN} \cdot m^2,$$
$$\bar{x}_2 = \frac{6 \text{ m}}{3} = 2 \text{ m}$$

$$\text{Area}_3 = \tfrac{1}{3}(3 \text{ m})(22.5 \text{ kN} \cdot m)$$
$$= 22.5 \text{ kN} \cdot m^2,$$
$$\bar{x}_3 = \frac{3 \text{ m}}{4} = 0.75 \text{ m}$$

$$d_B = \frac{\text{Area}_1 \bar{x}_1 - \text{Area}_2 \bar{x}_2 - \text{Area}_3 \bar{x}_3}{EI}$$

$$= \frac{304 \text{ kN} \cdot m^2 (3 \text{ m}) - 135 \text{ kN} \cdot m^2 (2 \text{ m})}{200 \text{ GPa}(7.12 \times 10^{-5} \text{ m}^4)}$$

$$- \frac{22.5 \text{ kN} \cdot m^2 (0.75 \text{ m})}{200 \text{ GPa}(7.12 \times 10^{-5} \text{ m}^4)}$$

$$= 0.0439 \text{ m}$$

STEP 5.

Determine θ_{AM} by proportionality.

$$\theta_{AM} \cong \tan\theta_A \cong \frac{d_{B/A}}{L} \cong \frac{1.61\ \text{in.}}{360\ \text{in.}}$$

$$\cong 0.00448\ \text{rad}$$

STEP 6.

Determine θ_{AM} by the first moment-area theorem.

$$\theta_{AM} = \frac{\text{Area}_{AM}}{EI}$$

remembering that Area_{AM} is the area above the $M = 0$ line minus the area below the $M = 0$ line to the left of M.

$$\theta_{AM} = \frac{\frac{1}{2}(1500\ \text{lb})(x)(x)}{29{,}000{,}000\ \text{psi}\ (171\ \text{in.}^4)}$$

$$- \frac{\frac{1}{2}(1500\ \text{lb})(x - 120\ \text{in.})(x - 120\ \text{in.})}{29{,}000{,}000\ \text{psi}\ (171\ \text{in.}^4)}$$

$$= 3.63 \times 10^{-5}(x) - 2.18 \times 10^{-3}$$

STEP 7.

Equate the Step 5 and Step 6 θ_{AM}s and solve for x.

$$3.63 \times 10^{-5}(x) - 2.18 \times 10^{-3} = 0.00448$$

$$x = 183\ \text{in.}$$

STEP 8.

Determine the maximum deflection (at $x = 183$ in.). This may be determined by determining the distance of point A from a tangent drawn to the curve at M.

STEP 5.

Determine θ_{AM} by proportionality.

$$\theta_{AM} \cong \tan\theta_A \cong \frac{d_B}{L} \cong \frac{0.0439\ m}{9\ m}$$

$$\cong 4.88 \times 10^{-3}\ rad$$

STEP 6.

Determine θ_{AM} by the first moment-area theorem.

$$\theta_{AM} = \frac{Area_{AM}}{EI}$$

remembering that $Area_{AM}$ is the area above the $M = 0$ line minus the area below the $M = 0$ line to the left of M.

$$\theta_{AM} = \frac{\frac{1}{2}(7500\ N)(x)(x)}{200\ GPa(7.12 \times 10^{-5}\ m^4)}$$

$$- \frac{\frac{1}{2}(7500\ N)(x - 3\ m)(x - 3\ m)}{200\ GPa(7.12 \times 10^{-5}\ m^4)}$$

$$= 1.58 \times 10^{-3}x - 2.37 \times 10^{-3}$$

STEP 7.

Equate the Step 5 and Step 6 θ_{AM} s and solve for x.

$$1.58 \times 10^{-3}\ x - 2.37 \times 10^{-3} = 4.88 \times 10^{-3}$$

$$x = 4.59\ m$$

STEP 8.

Determine the maximum deflection (at $x = 4.59$ m). This may be determined by determining the distance of point A from a tangent to the curve at M.

233

$$\Delta_M = \frac{\text{Area}_{AM}\bar{x}_{AM/A}}{EI} \quad \text{at } x = 183 \text{ in.}$$

where $\bar{x}_{AM/A}$ is the moment arm of Area_{AM} from a vertical axis through A.

$$\Delta_M = \frac{\frac{1}{2}(1500 \text{ lb})(183 \text{ in.})^2(2)\left(\left(\frac{183 \text{ in.}}{3}\right)\right)}{29{,}000{,}000 \text{ psi}\,(171 \text{ in.}^4)}$$

$$- \frac{\frac{1}{2}(1500 \text{ lb})(183 \text{ in.} - 120 \text{ in.})^2 183 \text{ in.}}{29{,}000{,}000 \text{ psi}\,(171 \text{ in.}^4)}$$

$$- \frac{\dfrac{183 \text{ in.} - 120 \text{ in.}}{3}\Big)}{29{,}000{,}000 \text{ psi}\,(171 \text{ in.}^4)}$$

$$= 0.521 \text{ in.}$$

$$\Delta_M = \frac{\text{Area}_{AM}\bar{x}_{AM/A}}{EI} \quad \textbf{at } x = 4.59 \, m$$

where $\bar{x}_{AM/A}$ is the moment arm of Area_{AM} from a vertical axis through A.

$$\Delta_M = \frac{\frac{1}{2}(7500 \, N)(4.59)^2(\frac{2}{3})(4.59)}{200 \, GPa\,(7.12 \times 10^{-5} \, m^4)}$$

$$- \frac{\frac{1}{2}(7500 \, N)(4.59 - 3)^2(3 + \frac{2}{3}(4.59 - 3))}{200 \, GPa\,(7.12 \times 10^{-5} \, m^4)}$$

$$= 0.0143 \, m$$

9-4   DEFLECTIONS BY SUPERPOSITION

When a beam carries more than one load, perhaps of different types, the deflection of the beam at any point is the sum of the deflections that would be caused at that point by each of the individual loads, provided that the beam is loaded within the range where it obeys Hooke's law (acts elastically). The principle just stated is an application of the *principle of superposition* to the determination of the deflection of beams. This provides us with a comparatively simple method for determining the deflection of beams with more than one load or different types of load. However, the point at which the deflection is desired must be known. The method of superposition is not suitable for use where the maximum deflection of a beam is required and that point is not obvious. In such cases the moment-area method should be used.

To use the method of superposition we need to know the deflection equations for a variety of types of loads. A number of these are given in Table 9-2. Many others are available in engineering handbooks and manufacturer's literature. To illustrate the use of this method, and at the same time demonstrate its validity, we will determine the deflection of the beam in the last example of the previous section. In that example we determined the deflection to be 0.521 in. at 183 in. from the left-hand end of the beam for the Customary version and 0.0143 m at 4.59 m from the left-hand end of the beam in the Metric version. We should be able to duplicate these deflections using the superposition method.

Table 9-2

Beam Deflection Equations

Δ	Deflection of vertical distance from a horizontal line passing through the support or supports of the beam on its neutral axis. Δ is positive downward.
Δx	Deflection of any point on the beam in the direction perpendicular to the x axis.
Δ_{max}	Maximum deflection.
θ	Angle the beam makes with a horizontal line, that is, the slope of the beam.

1.

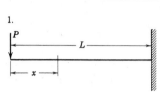

$$\Delta_x = \frac{P}{6\,EI}\,(2L^3 - 3L^2x + x^3)$$

$$\Delta_{max}\text{(at free end)} = \frac{PL^3}{3\,EI}$$

$$\theta_{\text{free end}} = \frac{PL^2}{2\,EI}$$

2.

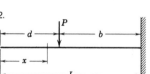

$$\Delta_x = \frac{P(L-x)^2}{6\,EI}\,(3b - L - x) \quad \text{for} \quad 0 \leqslant x \leqslant a$$

$$\Delta_x = \frac{Pb^2}{6\,EI}\,(3x - b) \quad \text{for} \quad a < x < L$$

$$\Delta_{max} = \frac{Pb^2}{6\,EI}\,(3L - b) \qquad \text{at free end}$$

$$\theta_{\text{free end}} = \frac{Pb^2}{2\,EI}$$

3.

q lb per unit length

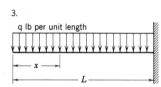

$$\Delta_x = \frac{q}{24\,EI}\,(x^4 - 4L^3x + 3L^4)$$

$$\Delta_{max} = \frac{qL^4}{8\,EI} \qquad \text{at free end}$$

$$\theta_{\text{free end}} = \frac{qL^3}{6\,EI}$$

4.

$$\Delta_x = \frac{M(L-x)^2}{2\,EI}$$

$$\Delta_{\text{max (at free end)}} = \frac{ML^2}{2\,EI}$$

$$\theta_{\text{free end}} = \frac{ML}{EI}$$

235

Table 9.2 (Contd.)

5.

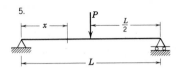

$$\Delta_x = \frac{Px}{48\,EI}\,(3L^2 - 4x^2) \quad \text{for} \quad 0 \leqslant x \leqslant \frac{L}{2}$$

$$\Delta_{max} = \frac{PL^3}{48\,EI} \quad \text{at center}$$

$$\theta_{at\ ends} = \frac{PL^2}{16\,EI}$$

6.

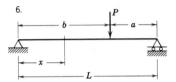

$$\Delta_x = \frac{Pax}{6\,LEI}\,(L^2 - x^2 - a^2) \quad \text{for} \quad 0 \leqslant x \leqslant b$$

$$\Delta_x = \frac{Pa}{6\,LEI}\left[\frac{L}{a}\,(x - b)^3 + (L^2 - a^2)x - x^3\right]$$
$$\text{for} \quad b < x < L$$

$$\Delta_{max} = \frac{Pa\,(L^2 - a^2)^{3/2}}{9\sqrt{3}\,LEI} \quad \text{at} \quad x = \sqrt{\frac{L^2 - a^2}{3}}$$

$$\theta_{left\ end} = \frac{Pa\,(L^2 - a^2)}{6\,LEI}, \quad \theta_{right\ end} = \frac{Pab\,(2L - a)}{6\,LEI}$$

7.

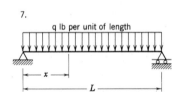

q lb per unit of length

$$\Delta_x = \frac{qx}{24\,EI}\,(L^3 - 2Lx^2 + x^3)$$

$$\Delta_{max} = \frac{5qL^4}{384\,EI} \quad \text{at center}$$

$$\theta_{at\ ends} = \frac{qL^3}{24\,EI}$$

8.

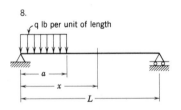

q lb per unit of length

$$\Delta_x = \frac{qx}{24\,EIL}\,[a^2(2L - a)^2 - 2ax^2(2L - a) + Lx^3]$$
$$\text{for} \quad 0 \leqslant x \leqslant a$$

$$\Delta_x = \frac{qa^2(L - x)}{24\,EIL}\,(4Lx - 2x^2 - a^2) \quad \text{for} \quad a \leqslant x \leqslant L$$

$$\theta_{left\ end} = \frac{qa^2}{24\,LEI}\,(a^2 - 4aL + 4L^2)$$

$$\theta_{right\ end} = \frac{qa^2}{24\,LEI}\,(2L^2 - a^2)$$

9.

$$\Delta_x = \frac{MLx}{6\,EI}\left(L - \frac{x^2}{L^2}\right)$$

$$\Delta_{max} = \frac{ML^2}{9\sqrt{3}EI} \quad \text{at} \quad x = \frac{L}{\sqrt{3}}$$

$$\theta_{left\ end} = \frac{ML}{6\,EI}, \quad \theta_{right\ end} = \frac{ML}{3\,EI}$$

Example

customary

Determine the deflection of the beam shown in Figure 9-5(a) at the point 183 in. from its left-hand end.

metric

Determine the deflection of the beam shown in Figure 9-6(a) at the point 4.59 m from its left-hand end.

Solution

customary

Concentrated Load—Deflection at 183 in. from left end. Use Case 6, Table 9-2.

$$\Delta_{x_c} = \frac{Pa}{6\,LEI}\left[\frac{L}{a}(x-b)^3 + (L^2-a^2)x - x^3\right]$$

$x = 183$ in.
$b = 120$ in.
$a = 240$ in.
$L = 360$ in.
$P = 1500$ lb
$E = 29 \times 10^6$ psi
$I = 171$ in.^{4}

$$\Delta_{x_c} = \frac{1500\,\text{lb}\,(240\,\text{in.})}{6(360\,\text{in.})(29\times10^6\,\text{psi})(171\,\text{in.}^4)}$$

$$\left[\frac{360\,\text{in.}}{240\,\text{in.}}(183\,\text{in.}-120\,\text{in.})^3\right.$$

$$+ [(360\,\text{in.})^2 - (240\,\text{in.})^2]183\,\text{in.}$$

$$\left. - (183\,\text{in.})^3\right]$$

$$= 0.249\,\text{in.}$$

Uniformly Distributed Load—Deflection at 183 in. from left end.

$$\Delta_{x_u} = \frac{qa^2(L-x)}{24\,LEI}(4\,Lx - 2x^2 - a^2)$$

$q = 300$ lb/ft $= 25$ lb/in.
$a = 120$ in.
$L = 360$ in.
$x = 360$ in. $- 183$ in. $= 177$ in.

metric

Concentrated Load—Deflection at 4.59 m from left end.

$$\Delta_{x_c} = \frac{Pa}{6\,LEI}\left[\frac{L}{a}(x-b)^3 + (L^2-a^2)x - x^3\right]$$

$x = 4.59$ m
$a = 6$ m
$P = 7500$ N
$I = 7.12 \times 10^{-5}$ m^4
$b = 3$ m
$L = 9$ m
$E = 20$ GPa

$$\Delta_{x_c} = \frac{7500\,N\,(6\,m)}{6(9\,m)(200\,GPa)(7.12\times10^{-5}\,m^4)}$$

$$\times\left[\frac{9\,m}{6\,m}(4.59\,m - 3\,m)^3\right.$$

$$+ [(9\,m)^2 - (6\,m)^2]4.59\,m - (4.59\,m)^3\Big]$$

$$= 6.78 \times 10^{-3}\,m$$

Uniformly distributed Load—Deflection at 4.59 m from left end.

$$\Delta_{x_u} = \frac{qa^2(L-x)}{24\,LEI}(4\,Lx - 2x^2 - a^2)$$

$q = 5000$ N/m
$a = 3$ m
$L = 9$ m
$x = 9$ m $- 4.59$ m $= 4.41$ m

Note: Distance x is from the right-hand (loaded) end.

$$\Delta_{x_u} = \frac{25 \text{ lb/in.} (120 \text{ in.})^2 (360 \text{ in.} - 177 \text{ in.})}{24 (360 \text{ in.}) (29 \times 10^6 \text{ psi}) (171 \text{ in.}^4)}$$

$$\times [4(360 \text{ in.})(177 \text{ in.}) - 2(177)^2 - (120 \text{ in.})^2]$$

$$= 0.273 \text{ in.}$$

$$\Delta_{183 \text{ in.}} = \Delta_{x_c} + \Delta_{x_u} = 0.249 \text{ in.} + 0.273 \text{ in.}$$

$$= 0.522 \text{ in.}$$

This compares with 0.521 determined by the moment-area method, the slight difference being due to the difference in rounding off values.

Note: *Distance x is from the right-hand (loaded) end.*

$$\Delta_{x_u} = \frac{5000 \text{ N/m} (3 \text{ m})^2 (9 \text{ m} - 4.41 \text{ m})}{24 (9 \text{ m}) (200 \text{ GPa}) (7.12 \times 10^{-5} \text{ m}^4)}$$

$$\times [4(9 \text{ m})(4.41) - 2(4.41 \text{ m})^2 - (3 \text{ m})^2]$$

$$= 7.44 \times 10^{-3}$$

$$\Delta = \Delta_{x_c} + \Delta_{x_u} = 6.78 \times 10^{-3} + 7.44 \times 10^{-3}$$

$$= 0.0142 \text{ m}$$

which compares with 0.0143 m determined by the moment-area method, the slight difference being due to the difference in rounding off values.

9-5   ACCOMPLISHMENT CHECKLIST

From your study of this chapter you should now be able to

1. **Understand the terms used in describing the deflection of beams.**

2. **Determine the deflections of beams by applying the moment-area theorems.**

3. **Determine the point of maximum deflection of a beam by applying the moment-area theorems.**

4. **Determine the deflection of beams at any given point by the method of superposition.**

9-6   SUMMARY

The shape of the neutral axis of a loaded beam is called its *elastic curve*. The radius of the curve of the elastic curve at any point is called its *radius of curvature* at that point. It is designated by the Greek letter rho ρ. The reciprocal of the radius of curvature is called the *curvature*. The product EI is commonly referred to as the *flexural rigidity* of a beam.

The *first moment-area theorem* may be stated—*The angle (in radians) between the tangents to the elastic curve at any two points A and B is equal to the area of the moment diagram of the beam between the ordinates of those same two points, divided by the flexural rigidity EI of the beam.*

The *second moment-area theorem* may be stated—*If a tangent is drawn to an elastic curve at any point A, the vertical distance of any other point B (on the elastic curve) from that tangent is the moment with respect to B of the area of the moment diagram between the ordinates of points A and B divided by the flexural rigidity EI.*

The *principle of superposition* as applied to the deflection of beams states that *the deflection of a beam carrying more than one load, perhaps of different types, will at any point be equal to the sum of the deflections that would be caused at that point by each of the loads acting individually.*

PROBLEMS

Solve the following problems by applying the moment-area theorems. Unless otherwise specified the weight of the beams is to be ignored. All bending is to be assumed to be in the vertical plane.

9-1C   The cantilever beam in the illustration supports a load P of 2000 lb at a distance of 12 ft from the supporting wall. Determine the slope of the beam at the load if E is 29×10^6 psi and I is 14.8 in.^{4}

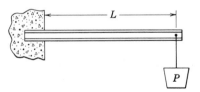

PROBLEM 9-1

9-1M   *The cantilever beam in the illustration supports a load P of 1000 kg at a distance L of 4 m from the supporting wall. Determine the slope of the beam at the load if E is 200 GPa and I is 6×10^{-6} m^4.*

9-2C   The figure shows a 1-15/16 diameter steel shaft mounted in two ball bearings which permit it to move endways and to bend. The load P is 8000 lb and L is

48 in. Determine the angle the shaft makes with the centerline of the bearings. E for steel is 29,000,000 psi.

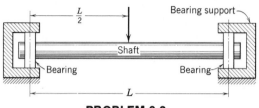

PROBLEM 9-2

9-2M *The figure shows a 50 mm steel shaft mounted in two ball bearings which permit it to move sideways and to bend. The load P is 4000 kg and L is 1.2 m. Determine the angle the shaft makes with the centerline of the bearings at the center of the bearings. E for steel is 200 GPa.*

9-3C A simple beam 20 ft long has concentrated loads of 10 kip 5 ft from each end. Determine the angle the beam makes with its supports if E is 29×10^6 psi and I is 204 in.^{4}

9-3M *A simple beam 6 m long has concentrated loads of 5000 kg 1.5 m from each end. Determine the angle the beam makes with its supports if E is 200 GPa and I is $8.5 \times 10^{-5} m^4$.*

9-4C A simple beam 20 ft long has a uniform load of 1 kip/ft. Determine the angle the beam makes with its supports if E is 29×10^6 psi and I is 204 in.^{4}

9-4M *A simple beam 6 m long carries a uniform load of 1500 kg/m. Determine the angle the beam makes with its supports if E is 200 GPa and I is $8.5 \times 10^{-5} m^4$.*

9-5C A cantilever beam 10 ft long carries a uniform load of 1000 lb/ft. Determine the slope of the end of the beam if E is 29×10^6 psi and I is 204 in.^{4}

9-5M *A cantilever beam 3 m long carries a uniform load of 1500 kg/m. Determine the slope of the end of the beam if E is 200 GPa and I is $8.5 \times 10^{-5} m^4$.*

9-6C   Determine the deflection of the beam of Problem 9-1C at the point of application of the load.

9-6M   *Determine the deflection of the beam of Problem 9-1M at the point of application of the load.*

9-7C   Determine the deflection of the shaft of Problem 9-2C at the point of application of the load.

9-7M   *Determine the deflection of the shaft of Problem 9-2M at the point of application of the load.*

9-8C   The 12 ft long cantilever beam in the illustration supports a load P of 2000 lb at a distance a of 8 ft from the supporting wall as shown in the figure. Determine the deflection of the end of the beam at the load if E is 29,000,000 psi and I is 44.8 in.^{4}

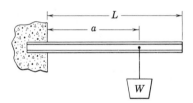

PROBLEM 9-8

9-8M   *The 7.5 m long cantilever beam in the illustration supports a load P of 1000 kg at a distance a of 5 m from the supporting wall. Determine the deflection of the beam at the load if E is 200 GPa and I is $1.86 \times 10^{-5}\,m^4$.*

9-9C   A concrete slab 15 ft long, 6 ft wide, and 6 in. thick is supported by two W10X21 wide-flanged beams to form a dock for small boats as shown in the figure. The concrete weighs 60 lb/ft^3 and E for the steel is 29×10^6 psi. Determine the deflection of the beams at the end of the dock including the effect of the weight of the beams.

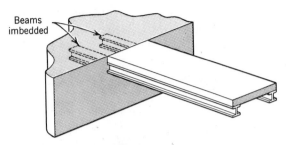

PROBLEM 9-9

9-9M *A concrete slab 4.5 m long, 1.8 m wide, and 15 cm thick is supported by two $250 \times 125 - 29.6$ wide-flanged beams to form a dock for small boats as shown in the figure. The concrete weighs $1000\,kg/m^3$ and E for the steel is 200 GPa. Determine the deflection of the beams at the end of the dock including the effect of the weight of the beams.*

9-10C Determine the maximum deflection of the wood floor joist shown if it supports a load of 27.5 lb/ft with $L = 24$ ft, $b = 4$ in., and $h = 8$ in. E for the wood is 1,500,000 psi.

PROBLEM 9-10

9-10M *Determine the maximum deflection of the wood floor joist shown if it supports a uniform load of $40\,kg/m$ with $L = 8\,m$, $b = 100\,mm$, and $h = 200\,mm$. E for the wood is 10.3 GPa.*

9-11C The figure shows a 1-15/16 diameter steel shaft mounted in two ball bearings which permit it to move endways and to bend. L is 48 in., a is 12 in., P is 800 lb, and E is 30×10^6 psi. Determine the slope of the shaft as it passes through the bearings and its maximum deflection.

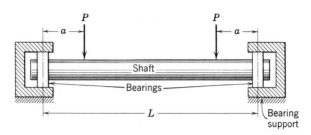

PROBLEM 9-11

9-11M *The figure shows a 50 mm diameter shaft mounted in two ball bearings which permit it to move endways and to bend. L is 1.2 m, a is 30 cm, P is 400 kg, and E is 207 GPa. Determine the slope of the shaft as it passes through the bearings and its maximum deflection.*

9-12C A W12X40 cantilever beam extends a distance $L = 12$ ft from the wall. Determine the concentrated load P applied at its free end, required to deflect the end of the beam 0.040 in. E is 29×10^6 psi.

9-12M *A $300 \times 200 - 65.4$ wide-flange cantilever beam extends a distance $L = 4$ m from the wall. Determine the concentrated load P applied at its free end, required to deflect the end of the beam 10 mm. E is 200 GPa.*

9-13C Determine the maximum uniform load a simple beam 3 in. wide by 12 in. high by 24 ft long could carry within a deflection limit of 1/360 of the length of the span if it is made of commercial white ash lumber having an E of 1.68×10^6 psi.

9-13M *Determine the maximum uniform load a simple beam 7.5 cm wide by 30 cm high by 7 m long could carry within a deflection limitation of 1/360 of the length of the span if it is made of commercial white ash having an E of 11.6 GPa.*

9-14C A W12X53 wide-flange beam 50 ft long is simply supported at its ends. Determine the deflection at the center of the beam due to its own weight. E is 29×10^6 psi.

9-14M   *A 300 × 300 − 84.5 wide-flange beam 15 m long is simply supported at its ends. Determine the deflection at the center of the beam due to its own weight. E is 200 GPa.*

9-15C   A cantilever beam extending 20 ft from its support has a uniform load of 275 lb/ft plus its own weight over its entire length and a 1000 lb concentrated load at its end. The beam is a W12X40 with an E of 29×10^6 psi. Determine its maximum deflection.

9-15M   *A cantilever beam extends 6 m from its support and has a uniform load of 400 kg plus its own weight over its entire length and a 500 kg concentrated load at its end. The beam is a 300 × 200 − 65.4 wide-flange with an E of 200 GPa. Determine its maximum deflection.*

9-16C   A W14X87 floor beam 50 ft long is designed to carry a uniformly distributed load over its entire length of 400 lb/ft including the weight of the beam and a concentrated 10,000 lb load at its center. Determine its maximum deflection. E is 29×10^6 psi.

9-16M   *A 350 × 350 − 131 wide-flange floor beam 15 m long is designed to carry a uniformly distributed load over its entire length of 600 kg/m including the weight of the beam and a 5 000 kg concentrated load at its center. Determine its maximum deflection. E is 200 GPa.*

9-17C   The cantilever beam in the figure carries a uniformly distributed load of 400 lb/ft and a concentrated load P of 10,000 lb. L is 18 ft, a is 12 ft, E is 29×10^6 psi, and I is 171 in.[4] Determine the deflection of the end of the beam.

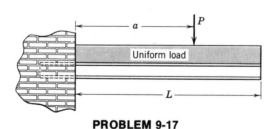

PROBLEM 9-17

9-17M   *The cantilever beam shown carries a uniformly distributed load of 600 kg/m and a concentrated load P of 4 500 kg. L is 5.5 m, a is 3.7 m, E is*

200 GPa, and I is $7.12 \times 10^{-5} m^4$. Determine the deflection of the end of the beam.

9-18C   A simply supported W12X53 wide-flange beam 20 ft long supports a single 8000 lb concentrated load 6 ft from its left-hand end. Determine the maximum deflection of the beam and locate the point at which it would occur. E is 29×10^6 psi.

9-18M   *A simply supported $300 \times 300 - 84.5$ wide-flange beam supports a single 4 000 kg concentrated load 2 m from its left-hand end. Determine the maximum deflection of the beam and locate the point at which it would occur. E is 200 GPa.*

9-19C   A simply supported W24X100 beam 26 ft long supports a uniform load of 1000 lb/ft extending for 9 ft toward its center from its right-hand end. Determine the location and amount of its maximum deflection if E is 29×10^6 psi.

9-19M   *A simply supported wide-flange beam 8 m long having an E of 200 GPa and an I of $1.25 \times 10^{-3} m^4$ supports a uniform load of 1 500 kg/m extending for 2.8 m toward its center from its right-hand end. Determine the location and amount of its maximum deflection.*

9-20C   A simply supported beam 26 ft long in the figure supports the concentrated 8000 lb load P and the 1000 lb/ft uniform load. Determine the location and magnitude of the maximum deflection of the beam if E is 29×10^6 psi, I is 3000 in.^{4}, a is 6 ft, and b is 9 ft.

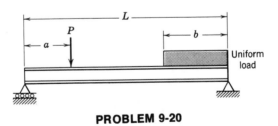

PROBLEM 9-20

9-20M   *The simply supported beam in the figure is 8 m long and supports the 36 kN load P and the uniform load which is 1 500 kg/m. E is 200 GPa, I is*

$1.25 \times 10^{-3} m^4$, a is 1.8 m, and b is 2.8 m. Determine the location and magnitude of the maximum deflection.

Solve the following problems by the superposition method. Unless otherwise specified the weight of the beams is to be ignored. All bending is assumed to be in the vertical plane.

9-21C   Determine the deflection of the free end of the beam in Problem 9-8C.

9-21M   *Determine the deflection of the free end of the beam in Problem 9-8M.*

9-22C   Determine the deflection of the beams at the end of the dock in Problem 9-9C.

9-22M   *Determine the deflection of the beams at the end of the dock in Problem 9-9M.*

9-23C   Determine the maximum deflection of the wood floor joist of Problem 9-10C.

9-23M   *Determine the maximum deflection of the wood floor joist of Problem 9-10M.*

9-24C   Determine the maximum deflection of the shaft in Problem 9-11C.

9-24M   *Determine the maximum deflection of the shaft in Problem 9-11M.*

9-25C   Determine the required concentrated load P in Problem 9-12C.

9-25M   *Determine the required concentrated load P in Problem 9-12M.*

9-26C   Determine the maximum required uniform load for the simple beam in Problem 9-13C.

9-26M   *Determine the required maximum uniform load for the simple beam in Problem 9-13M.*

9-27C   Determine the deflection at the center of the beam in Problem 9-14C due to its own weight.

9-27M   *Determine the deflection at the center of the beam in Problem 9-14M due to its own weight.*

9-28C   Determine the maximum deflection of the beam in Problem 9-15C.

9-28M   *Determine the maximum deflection of the beam in Problem 9-15M.*

9-29C   Determine the maximum deflection of the beam in Problem 9-16C.

9-29M   *Determine the maximum deflection of the beam in Problem 9-16M.*

9-30C   Determine the maximum deflection of the beam in Problem 9-17C.

9-30M   *Determine the maximum deflection of the beam in Problem 9-17M.*

9-31 through 9-33. Determine the deflection at the end of the beam shown in the figure.

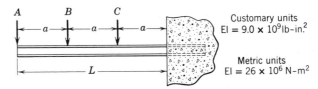

PROBLEMS 9-31 to 9-33

9-31C   A   10 kip   B   7 kip   C   0   L   15 ft

9-31M   *A   4 500 kg   B   3 000 kg   C   0   L   4.5 m*

9-32C   A   10 kip   B   10 kip   C   10 kip   L   18 ft

9-32M   *A   4 500 kg   B   4 500 kg   C   4 500 kg   L   6 m*

DEFLECTION OF BEAMS

9-33C   A − 5 kip   B 10 kip   C 5 kip   L 24 ft

9-33M   A − 3 000 kg   B 4 500 kg   C 3 000 kg   L 8 m

9-34 through 9-37. Determine the deflection at the end of the beam shown in the figure.

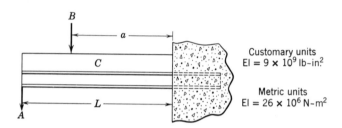

PROBLEMS 9-34 to 9-37

9-34C   A 10 kip   B 10 kip   C 300 lb/ft   L 10 ft   a 5 ft

9-34M   A 4 500 kg   B 4 500 kg   C 450 kg/m   L 4 m   a 2 m

9-35C   A 5 kip   B 20 kip   C 150 lb/ft   L 10 ft   a 8 ft

9-35M   A 2 000 kg   B 9 000 kg   C 200 kg/m   L 4 m   a 3 m

9-36C   A 20 kip   B 5 kip   C 300 lb/ft   L 10 ft   a 3 ft

9-36M   A 9 000 kg   B 3 000 kg   C 450 kg/m   L 3 m   a 1 m

9-37C   A − 10 kip   B 0   C 300 lb/ft   L 10 ft   a 5 ft

9-37M   A − 4 500 kg   B 0   C 450 kg/m   L 4 m   a 2 m

9-38 through 9-41. Determine the deflection at the center of the beam in the figure.

248

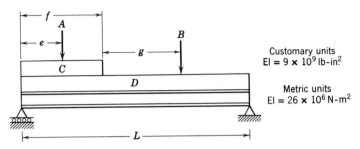

PROBLEMS 9-38 to 9-41

9-38C $A = 10$ kip, $B = 0$, $C = 0$, $D = 200$ lb/ft, $e = 5$ ft, $L = 30$ ft.

9-38M $A = 5$ Mg, $B = 0$, $C = 0$, $D = 300$ kg/m, $e = 2$ m, $L = 12$ m.

9-39C $A = 10$ kip, $B = 10$ kip, $C = 300$ lb/ft, $D = 0$, $e = 5$ ft, $f = 10$ ft, $g = 10$ ft, $L = 30$ ft.

9-39M $A = 5$ Mg, $B = 5$ Mg, $C = 450$ kg/m, $D = 0$, $e = 2$ m, $f = 4$ m, $g = 4$ m, $L = 12$ m.

9-40C $A = 10$ kip, $B = 10$ kip, $C = 300$ lb/ft, $D = 200$ lb/ft, $e = 5$ ft, $f = 10$ ft, $g = 10$ ft, $L = 30$ ft.

9-40M $A = 5$ Mg, $B = 5$ Mg, $C = 450$ kg/m, $D = 300$ kg/m, $e = 1.5$ m, $f = 3$ m, $g = 3$ m, $L = 9$ m.

9-41C $A = 10$ kip, $B = 10$ kip, $C = 300$ lb/ft, $D = 200$ lb/ft, $e = 5$ ft, $f = 10$ ft, $g = 10$ ft, $L = 30$ ft. Include 40 lb/ft weight of the beam in the determination of the deflection.

9-41M $A = 5$ Mg, $B = 5$ Mg, $C = 450$ kg/m, $D = 300$ kg/m, $e = 1.5$ m, $f = 3$ m, $g = 3$ m, $L = 9$ m. Include 60 kg/m weight of the beam in the determination of the deflection.

CHAPTER
OBJECTIVES ————————————

The object of your studies of this chapter is to learn how to

1. **Identify a statically indeterminate beam.**

2. **Determine the reactions of a statically indeterminate beam by the superposition method.**

3. **Determine the reactions of a statically indeterminate beam by the moment-area method.**

4. **Determine the reactions of a statically indeterminate continuous beam by a combination of the moment-area and superposition methods.**

CHAPTER ——————————10
STATICALLY INDETERMINATE
BEAMS

The last two chapters have been devoted to the determination of the stresses in and deflections of statically determinate beams under various kinds of loading. All of the unknown reactions of statically determinate beams may be determined by the application of one or more of the three fundamental equations of statics—$\Sigma M = 0$, $\Sigma H = 0$, and $\Sigma V = 0$. There are many combinations of beam supports and loads where the number of unknown reactions at the supports of a beam is more than the number of applicable equations from statics. Beams with such combinations of loads and supports are called *statically indeterminate beams.*

To determine the reactions of the supports of a statically indeterminate beam we must have as many mathematically distinct equations as we have unknowns. Thus, since the number of equations from statics is not enough, we must use equations based on the elastic qualities of the beams as developed in the previous chapter on beam deflections. Once we have determined all of the reactions of the supports of a beam we can determine its stresses and deflections by the methods of the previous two chapters.

A statically indeterminate beam has one or more reactions which can be removed and still leave the beam supported. Such reactions are called *redundant reactions.* The removal of the redundant reactions will leave the beam supported in a statically determinate fashion. The *measure of redundancy* is the number of reactions (including moments at supports) that must be removed to make the beam statically determinate. If only one reaction need be removed, the beam is said to be *statically indeterminate in the first degree*, if two, it is *statically indeterminate in the second degree*, and so on. Fortunately there will always be as many relationships based on elasticity available as there are degrees of indeterminancy.

10-1 SUPERPOSITION METHOD

The easiest method to use in determining the unknown reactions of a statically indeterminate beam is the method of superposition. In applying this method we take advantage of the fact that the total deflection of any point of a beam is equal to the sum of the deflections of that point caused by each of the individual forces and reactions acting separately. A case of special interest is where a redundant support keeps a point on a beam from deflecting. Here the sum of the deflections is zero. The following examples will demonstrate how we take advantage of this.

Example

Determine the reactions of the cantilever beam in Figure 10-1.

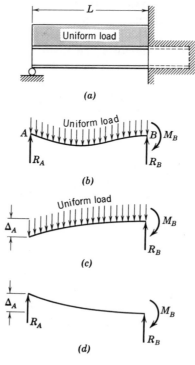

FIGURE 10-1

Solution

This beam is called a *propped cantilever beam*. Figure 10-1(b) is the free-body diagram of the beam drawn schematically with the elastic curve as imagined.

The beam is statically indeterminate in the first degree. We must remove one redundant reaction in order to perform our analysis. The reaction R_A is redundant. If it were removed, the beam would act as an ordinary cantilever beam and deflect as shown in Figure 10-1(c). For the end of the beam to remain fixed (the sum of the deflections equal to zero), the deflection that would be caused by the reaction R_A in the absence of the uniform load would have to be equal and opposite to that caused by the uniform load in the absence of R_A as in Figure 10-1(d). From Table 9-2 the deflection at end of the beam due to the uniform load $= qL^4/8\,EI$. Deflection at end of the beam due to $R_A = R_A L^3/3\,EI$. Equating the above deflections

$$\frac{R_A L^3}{3\,EI} = \frac{qL^4}{8\,EI}$$

Solving for R_A we find

$$R_A = \frac{3\,qL}{8}$$

Then by summation of vertical forces

$$\frac{3\,qL}{8} + R_B - qL = 0$$

$$R_B = \frac{5\,qL}{8}$$

Finally, we can determine M_B by summing moments about B.

$$M_B + R_A L - \frac{qL^2}{2} = 0$$

$$M_B = \frac{qL^2}{2} - R_A L = \frac{qL^2}{2} - \frac{3\,qL}{8}(L)$$

$$M_B = \frac{qL^2}{8}$$

Thus we have determined all of the unknown reactions and, knowing these, can proceed to determine the stresses and deflections at any point in the beam by the methods described in the previous two chapters.

Alternate Solution

In the above solution we removed the prop as being the redundant member. If we keep the prop in place, we can remove the rotational restraint of the wall permitting the beam to rotate at the right-hand end. Note in Figure 10-2(c) that the beam now becomes a simple beam. The bending of the beam will now cause the beam to have a downward (toward the left) slope at what was the fixed end. Since the slope at the wall was zero before removing the moment M, the effect of the moment must have tended to cause the beam to have an upward slope as in Figure 10-2(d) equal to the downward slope it would have as a simple beam; i.e., the sum of the slopes caused by the two loads acting separately will be zero. Slope due to uniform load acting on the simple

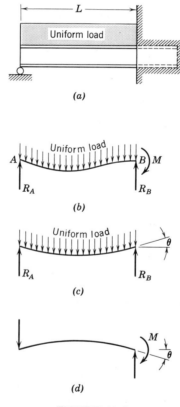

(a)

(b)

(c)

(d)

FIGURE 10-2

beam $= qL^3/24\,EI$. Slope due to moment acting at right-hand end of beam $=$ $ML/3\,EI$. Equate the slopes

$$\frac{ML}{3\,EI} = \frac{qL^3}{24\,EI}$$

Solve for M

$$M = \frac{qL^2}{8}$$

The equations of statics can now be used to determine the other unknown reactions.

Figure 10-3 shows a beam having three supports. It is statically indeterminate in the first degree. It can be readily analyzed by the application of the principle of superposition. First we remove the redundant support at the middle and determine the deflection caused by the uniform and concentrated loads without it. With the middle support in place the beam has a zero deflection at that support.

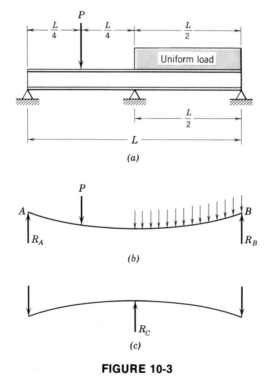

FIGURE 10-3

Therefore, to make this true, the deflection caused by the middle support acting upward in the absence of the loads must be equal to that caused by the loads acting downward. This procedure is demonstrated by the following examples.

Example

customary

Determine the reactions at the supports of the beam in Figure 10-3 by the superposition method when $L = 20$ ft, $q = 300$ lb/ft, and $P = 10,000$ lb.

metric

Determine the reactions at the supports of the beam in Figure 10-3 by the superposition method when $L = 8\,m$, $q = 300\,kg/m$, and $P = 4000\,kg$.

Solution

customary

Deflection at center of beam due to P with R_C removed.
From Case 6, Table 9-2,

$$\Delta_{x_P} = \frac{Pax}{6\,LEI}(L^2 - x^2 - a^2)$$

Substitute 5 ft for a, 10 ft for x, and 20 ft for L in the above equation.

$$\Delta_{x_P} = \frac{10,000\ \text{lb}\ (5\ \text{ft})(10\ \text{ft})}{6\,(20\ \text{ft})\,EI}$$
$$\times ((20\ \text{ft})^2 - (10\ \text{ft})^2 - (5\ \text{ft})^2)$$
$$= \frac{1375 \times 10^5}{120\,EI}\,\text{ft}$$

Deflection at center of beam due to uniform load with R_C removed.
From Case 8, Table 9-2,

$$\Delta_{x_q} = \frac{qx}{24\,EIL}(a^2(2L - a)^2$$
$$- 2ax^2(2L - a) + Lx^3)$$

metric

Deflection at center of beam due to P with middle support removed. From Case 6, Table 9-2,

$$\Delta_{x_P} = \frac{Pax}{6\,LEI}(L^2 - x^2 - a^2)$$

Substitute $2\,m$ for a, $4\,m$ for x, $8\,m$ for L, and $4000\,kg$ $(9.81\,N/kg)$ $39,200\,N$ for P.

$$\Delta_{x_P} = \frac{39,200\ N\,(2\,m)\,(4\,m)}{6(8\,m)\,EI}\,((8\,m)^2$$
$$- (4\,m)^2 - 2\,m)^2)$$
$$= \frac{138 \times 10^5}{48\,EI}\,m$$

Deflection at center of beam due to uniform load with center support removed.
From Case 8, Table 9-2,

$$\Delta_{x_q} = \frac{qx}{24\,EIL}(a^2(2L - a)^2$$
$$- 2ax^2(2L - a) + Lx^3)$$

Substitute 10 ft for a, 10 ft for x, 20 ft for L, and 300 lb/ft for q in the above equation.

$$\Delta_{x_q} = \frac{300 \text{ lb/ft} (10 \text{ ft})}{24 (20 \text{ ft}) EI} ((10 \text{ ft})^2 (40 \text{ ft} - 10 \text{ ft})^2$$

$$- 2 (10 \text{ ft})(10 \text{ ft})^2 (40 \text{ ft} - 10 \text{ ft})$$

$$+ 20 \text{ ft} (10 \text{ ft})^3)$$

$$= \frac{1500 \times 10^5}{480 \, EI} \text{ft}$$

$$\Delta_{x_P} + \Delta_{x_q} = \frac{1375 \times 10^5}{120 \, EI} \text{ft} + \frac{1500 \times 10^5}{480 \, EI} \text{ft}$$

$$= \frac{700 \times 10^5}{48 \, EI} \text{ft}$$

Deflection upward at center of beam due to R_C in the absence of the uniform and concentrated loads.

From Case 5, Table 9-2,

$$\Delta_{x R_C} = \frac{R_C L^3}{48 \, EI} = \frac{R_C (20 \text{ ft})^3}{48 \, EI}$$

$$= \frac{8000 \, R_C}{48 \, EI} \text{ft}$$

Since the total deflection of the beam is zero, the total downward deflection must equal the total upward deflection.

$$\Delta_{x_P} + \Delta_{x_q} = \Delta_{x R_C}$$

$$\frac{700 \times 10^5}{48 \, EI} \text{ft} = \frac{8000 \, R_C}{48 \, EI} \text{ft}$$

Solving for R_C gives us $R_C = 8750$ lb.

The other reactions can now be found by simple summing of moments.

Substitute 4 m for a, 4 m for x, 8 m for L, and 300 kg (9.81 N/kg)/m = 2940 N/m for q in the above equation.

$$\Delta_{x_q} = \frac{2940 \text{ N/m} (4 \text{ m})}{24 (8 \text{ m}) EI} (4 \text{ m})^2 (16 \text{ m} - 4 \text{ m})^2$$

$$- 2 (4 \text{ m})(4 \text{ m})^2 (16 \text{ m} - 4 \text{ m})$$

$$- 8 \text{ m}(4 \text{ m})^3)$$

$$= \frac{7.53 \times 10^5}{48 \, EI} \text{m}$$

The total downward deflection is

$$\Delta_{x_P} + \Delta_{x_P} = \frac{138 \times 10^5}{48 \, EI} \text{m} + \frac{7.53 \times 10^5}{48 \, EI} \text{m}$$

$$= \frac{146 \times 10^5}{48 \, EI} \text{m}$$

Deflection upward at center of beam due to R_C in the absence of the uniform and concentrated loads.

From Case 5, Table 9-2,

$$\Delta_{x R_C} = \frac{R_C L^3}{48 \, EI} = \frac{R_A (8 \text{ m})^3}{48 \, EI} = \frac{512 R_C}{48 \, EI} \text{m}$$

Since the total deflection of the beam is zero, the total downward deflection must equal the total upward deflection.

$$\Delta_{x_P} + \Delta_{x_q} = \Delta_{x R_C}$$

$$\frac{146 \times 10^5}{48 \, EI} \text{m} = \frac{512 R_C}{48 \, EI} \text{m}$$

Solving for R_C gives us $R_C = 28\,500$ N.

The other reactions can now be found by simple summing of moments.

10-2 MOMENT-AREA METHOD

The moment-area theorems enable us to determine the angle between the tangents of two points on a deflection curve and the vertical distance between a point on the deflection curve and the tangent drawn through another point on the curve. The equations based on elasticity that are needed to determine the reactions of statically indeterminate beams can be found by use of the two moment-area theorems. The conditions to look for are where two points on a curve are at the same level, in which case the sum of the moments of the moment diagram areas is zero, where two points on a curve have zero slope, in which case the area of the moment diagram between the points must equal zero, or where there are two equal slopes. Experience will develop your ability to spot these conditions and apply them.

Example

The beam in Figure 10-4 is built-in (fixed) at both ends. The free-body diagram, Figure 10-4(b), reveals four unknown reactions. Because there are only two static equilibrium equations available to us (there are no horizontal forces) it would seem that we would need to develop two equations based on elasticity. There is a surprise in store for you.

Solution

The beam is symmetrical. Consequently $R_A = R_B$ and $M_A = M_B$. By equilibrium of vertical forces

$$R_A = R_B = \frac{qL}{2}$$

We can now draw the moment diagram (by parts) as shown in Figure 10-4(c) where the plain shaded area is the moment-area of R_A, the vertically barred area is the moment-area of M_A, and the diagonally barred area is the moment-area of the uniform load.

At points A and B the elastic curve is horizontal, which means that the angle between the tangents at those points is zero. By the first moment-area theorem the area of the moment diagram between points A and B must therefore be zero.

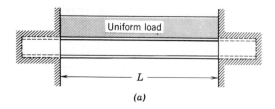

(a)

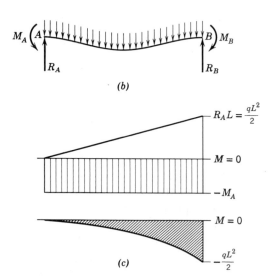

(b)

(c)

FIGURE 10-4

$$\frac{1}{2}\left(\frac{qL^2}{2}\right)(L) - M_A L - \frac{1}{3}L\left(\frac{qL^2}{2}\right) = 0$$

Solving for M_A we get

$$M_A = \frac{qL^2}{12}$$

Since M_B equals M_A we have determined the value of all of the unknown reactions.

The unknown reactions cannot always be so handily determined. The following example will illustrate.

Example

In Figure 10-5 we have a beam with both ends built in with a concentrated load P as shown. The free-body diagram showing the elastic curve of the beam reveals that we have four unknown reactions. Unlike the previous example, this is not a symmetrical system. No two reactions are the same.

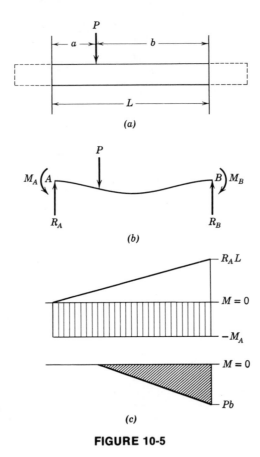

(a)

(b)

(c)

FIGURE 10-5

Solution

As in the previous example the tangents at A and B being horizontal yield one equation by the first moment-area theorem. However, this beam is statically indeterminate in the second degree so we need a second equation based on elasticity. It is available by the use of the second area-moment theorem. Since both A and B are at the same level the sum of the moments of the area of the moment diagram must equal zero. Solving these two moment area equations simultaneously will yield the values we need.

The moment diagram is shown in Figure 10-5(c). Using the first area-moment theorem we find,

$$\frac{R_A L^2}{2} - M_A L - \frac{Pb^2}{2} = 0 \tag{1}$$

Using the second moment-area theorem we find

$$\frac{R_A L^3}{6} - \frac{M_A L^2}{2} - \frac{Pb^3}{6} = 0 \tag{2}$$

Multiplying Equation (1) above by $L/3$ gives us

$$\frac{R_A L^3}{6} - \frac{M_A L^2}{3} - \frac{Pb^2 L}{6} = 0 \tag{3}$$

Solving Equations (2) and (3) simultaneously yields

$$M_A = \frac{Pab^2}{L^2}$$

Substituting this value of M_A in either Equation (1) or (2), we will find

$$R_A = \frac{2\,Pab^2}{L^3} + \frac{Pb^2}{L^2}$$

Since $\Sigma V = 0$,

$$R_B = P - R_A = P - \left(\frac{2\,Pab^2}{L^3} + \frac{Pb^2}{L^2}\right)$$

which can be reduced to

$$R_B = \frac{2\,Pa^2 b}{L^3} + \frac{Pa^2}{L^2}$$

The only remaining reaction to be determined is M_B. It can be determined by taking moments about B (or any other point for that matter).

$$\sum M_B = 0$$

$$-M_A + R_A L - Pb + M_B = 0$$

$$M_B = M_A - R_A L + Pb$$

$$= \frac{Pab^2}{L^2} - \frac{2\,Pab^2}{L^2} - \frac{Pb^2}{L} + Pb$$

which reduces to

$$M_B = \frac{Pa^2 b}{L^2}$$

An interesting application of the moment-area method is in the determination of the reactions of continuous beams such as that shown in Figure 10-6. Here

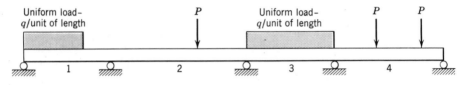

FIGURE 10-6

we have three redundant supports and hence would need three equations based on elasticity in addition to those of static equilibrium. These may be found by analyzing adjacent pairs of spans; spans 1 and 2, spans 2 and 3, and then spans 3 and 4. To demonstrate this we will examine the two middle spans as shown in Figure 10-7(a).

Figure 10-7(b) shows how we might imagine the deflection curve—greatly exaggerated and with all of the forces, loads, and moments that keep this section of the beam in equilibrium. The curve of a continuous beam over any support is smooth and continuous. In the figure a tangent has been drawn to the curve at B giving us the equal angles θ_1 and θ_2.

The next step is to separate the beam at B giving us the two simple beams in Figure 10-7(c). The middle reaction R_B has been divided into R_{B_1}, that part of the total reaction attributable to the left-hand span, and R_{B_2}, that attributable to the right-hand span. From above, $\theta_1 = \theta_2$ therefore $\tan \theta_1 = \tan \theta_2$, and

$$\frac{d_A}{L_1} = -\frac{d_B}{L_2}$$

the minus sign showing that d_A and d_B are deflections in opposite directions.

We can now determine d_A and d_B by applying the second moment-area theorem. To do this we must first draw the moment diagrams. Here we use

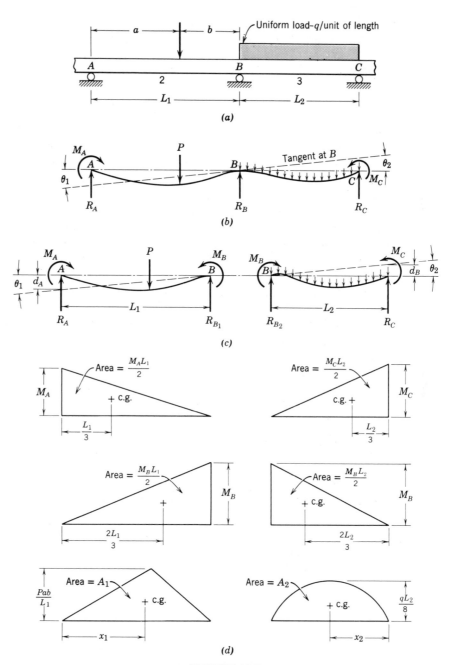

(a)

(b)

(c)

(d)

FIGURE 10-7

the principle of superposition and draw the moment diagrams in parts with a separate diagram being drawn for each moment and load acting on the beam segments. The top left diagram is the moment diagram of the left-hand span with the moment M_A only acting on it. Next below is the moment diagram of the span as it would be with only M_B acting on it. Finally, at the bottom is the moment diagram of the beam with only the concentrated load P acting on it. Likewise the diagrams for the right-hand span are drawn. We can now develop equations for d_A/L_1 and d_B/L_2 using the second moment-area theorem.

$$\frac{d_A}{L_1} = \frac{1/2\, M_A L_1 (L_1/3)}{EI\,(L_1)} + \frac{1/2\, M_B L_1 (2\,L_1/3)}{EI\,(L_1)} + \frac{A_1 \bar{x}_1}{EI\,(L_1)}$$

$$= \frac{M_A L_1}{6\,EI} + \frac{M_B L_1}{3\,EI} + \frac{A_1 \bar{x}_1}{L_1 EI} \qquad (a)$$

$$-\frac{d_B}{L_2} = -\left(\frac{1/2\, M_B L_2 (2\,L_2/3)}{EI\,(L_2)} + \frac{1/2\, M_C L_2 (L_2/3)}{EI\,(L_2)} + \frac{A_2 \bar{x}_2}{EI\,(L_2)}\right)$$

$$= -\frac{M_B L_2}{3\,EI} - \frac{M_C L_2}{6\,EI} - \frac{A_2 \bar{x}_2}{L_2 EI} \qquad (b)$$

Equating (a) and (b) and dropping the EIs because they appear in every term, we get

$$\frac{M_A L_1}{6} + \frac{M_B L_1}{3} + \frac{A_1 \bar{x}_1}{L_1} = -\frac{M_B L_2}{3} - \frac{M_C L_2}{6} - \frac{A_2 \bar{x}_2}{L_2}$$

Multiplying by 6, combining the M_B terms and transferring those terms *not* containing the moments M_A, M_B, or M_C to the right side of the equal sign gives us

$$M_A L_1 + 2 M_B (L_1 + L_2) + M_C L_2 = -\frac{6\,A_1 \bar{x}_1}{L_1} - \frac{6\,A_2 \bar{x}_2}{L_2} \qquad (10\text{-}1)$$

where M_A, M_B, and M_C are the bending moments at supports A, B, and C, respectively; L_1 is the length of the span between A and B, and L_2 is the length of the span between B and C. A_1 and A_2 are the areas of the moment diagrams of the loads. They are drawn as if each span of the beam were simply supported; $\bar{x}_1$ and $\bar{x}_2$ are the distances of the centroids of the moment diagrams from A and C, respectively. The terms on the right-hand side of the equation are sometimes called the *load terms*. The equation is known as the *three-moment equation*. The following examples will demonstrate the application of the three-moment equation.

Example

customary	metric
Determine the reactions at the supports of the beam in Figure 10-8(a).	*Determine the reactions at the sup- ports of the beam in Figure 10-9(a).*

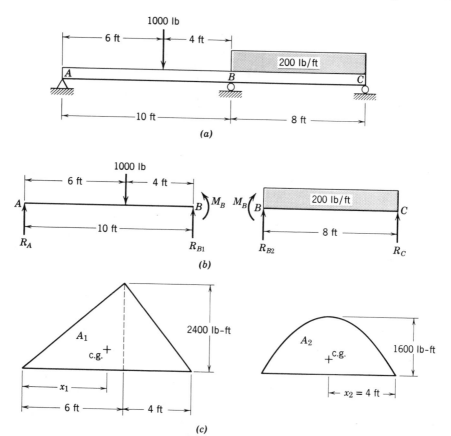

FIGURE 10-8

Solution

customary	metric
In Figure 10-8(b) are drawn the free-body diagrams of the left-hand span AB and	*In Figure 10-9(b) are drawn the free- body diagrams of the left-hand span AB,*

265

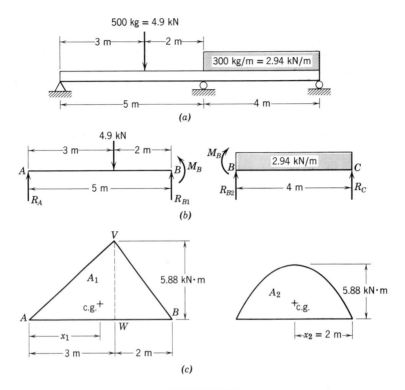

FIGURE 10-9

the right-hand span BC, of the beam. M_B is the bending moment of the beam at the middle support B, shown in the positive direction (tending to cause compression of the top fibers of the beam). R_{B1} represents that portion of the reaction of the support at B caused by the load on the span AB, and R_{B2} represents that caused by the load on span BC.

The moment diagrams in Figure 10-8(c) represent the effect of the loads on the spans as they would be if the spans were simply supported. There are no moments at A and C because the beam ends and is free to rotate at these points. The three-

and the right-hand span BC of the beam. M_B is the bending moment of the beam at the middle support B, shown in the positive direction (tending to cause compression of the top fibers of the beam). R_{B_1} represents that portion of the reaction of the support at B caused by the load on span AB and R_{B2} represents that caused by the load on span BC.

The moment diagrams in Figure 10-9(c) represent the effect of the loads on the spans as they would be if the spans were simply supported. There are no moments at A and C because the beam

moment equation is, therefore,

$$0 + 2M_B(L_1 + L_2) + 0 = -\frac{6A_1\bar{x}_1}{L_1} - \frac{6A_2\bar{x}_2}{L_2}$$

(a)

$A_1\bar{x}_1$ is the sum of the moments of areas AVW and WVB with respect to A.

$$A_1\bar{x}_1 = \tfrac{1}{2}(6 \text{ ft})(2400 \text{ lb-ft})(4 \text{ ft})$$

$$+ \tfrac{1}{2}(4 \text{ ft})(2400 \text{ lb-ft})(6 + \tfrac{4}{3} \text{ ft})$$

$$= 64,000 \text{ lb-ft}^3$$

$$-\frac{6A_1\bar{x}_1}{L_1} = -\frac{6(64,000 \text{ lb-ft}^3)}{10 \text{ ft}}$$

$$= -38,400 \text{ lb-ft}^2$$

The area of parabola A_2 is $2/3$ bh so

$$-\frac{6A_2\bar{x}_2}{L_2} = -\frac{6(\tfrac{2}{3})(8 \text{ ft})(1600 \text{ lb-ft})(4 \text{ ft})}{8 \text{ ft}}$$

$$= -25,600 \text{ lb-ft}^2$$

Substituting the known values in the three-moment Equation (a) yields

$$2M_B(10 \text{ ft} - 8 \text{ ft}) = -38,400 \text{ lb-ft}^2$$
$$- 25,600 \text{ lb-ft}^2$$

$$M_B = \frac{-64,000 \text{ lb-ft}^2}{36 \text{ ft}} = -1780 \text{ lb-ft}$$

The negative moment tells us that there is tension in the top fibers of the beam at point B.

Knowing M_B we can now determine reaction R_A by summing moments about R_{B1} of the left-hand span of the beam (Fig. 10-8(b)) and by summing moments about R_{B2} of the right-hand span we can find R_C.

ends and is free to rotate at these points. The three-moment equation can therefore be written

$$0 + 2M_B(L_1 + L_2) + 0 = -\frac{6A_1\bar{x}_1}{L_1} - \frac{6A_2\bar{x}_2}{L_2}$$

(a)

$A_1\bar{x}_1$ is the sum of the moments of areas AVW and WVB with respect to A.

$$A_1\bar{x}_1 = \tfrac{1}{2}(3 \text{ m})(5.88 \text{ kN·m})(2 \text{ m})$$

$$+ \tfrac{1}{2}(2 \text{ m})(5.88 \text{ kN·m})\left(3 \text{ m} + \frac{2 \text{ m}}{3}\right)$$

$$= 39.2 \text{ kN·m}^3$$

$$-\frac{6A_1\bar{x}_1}{L_1} = -\frac{6(39.2 \text{ kN·m}^3)}{5 \text{ m}}$$

$$= -47 \text{ kN·m}^2$$

The area of parabola A_2 is two-thirds the length of the diagram times its length, so

$$-\frac{6A_2\bar{x}_2}{L_2} = -\frac{6(2/3)(4 \text{ m})(5.88 \text{ kN·m})(2 \text{ m})}{4 \text{ m}}$$

$$= -47 \text{ kN·m}^2$$

Substituting these values in Equation (a) yields

$$2M_B(5 \text{ m} + 4 \text{ m}) + (-47 \text{ kN·m}^2)$$
$$- 47 \text{ kN·m}^2$$

$$M_B = \frac{-94 \text{ kN·m}^2}{18 \text{ m}} = -5.22 \text{ kN·m}$$

The negative moment tells us that there is tension in the top fibers of the beam at point B.

Knowing M_B we can now determine R_A by summing moments about R_{B1} of the left-hand span of the beam (Figure 10-9(b)) and by summing moments about R_{B2} of the right-hand span of the beam we can find R_C.

$$\sum M_{R_{B1}} = R_A(10\text{ ft}) - 1000\text{ lb}(4\text{ ft})$$
$$- (-1780\text{ lb-ft}) = 0$$
$$R_A = 222\text{ lb}$$

$$\sum M_{R_{B2}} = -R_C(8\text{ ft}) + 200\text{ lb/ft}(8\text{ ft})^2/2$$
$$+ (-1780\text{ lb-ft}) = 0$$
$$R_C = 578\text{ lb}$$

We can now determine R_B by the summation of vertical forces.

$$\sum V = 0$$

$$R_B = 1000\text{ lb} + 200\text{ lb/ft}(8\text{ ft}) - 222\text{ lb}$$
$$- 578\text{ lb}$$
$$= 1800\text{ lb}$$

$$\sum M_{R_{B1}} = R_A(5\text{ m}) - 4.9\text{ kN}(2\text{ m})$$
$$- (-5.22\text{ kN·m}) = 0$$
$$R_A = 0.916\text{ kN}$$

$$\sum M_{R_{B2}} = -R_C(4\text{ m}) + 2.94\text{ kN/m}(4\text{ m})^2/2$$
$$+ (-5.22\text{ kN · m})$$
$$R_C = 4.58\text{ kN}$$

We can now determine R_B by summation of the vertical forces on the entire beam.

$$\sum V = 0$$

$$R_B = 4.9\text{ kN} + 2.94\text{ kN/m}(4\text{ m})$$
$$- 0.916\text{ kN} - 4.58\text{ kN}$$
$$= 11.2\text{ kN}$$

10-3 ACCOMPLISHMENT CHECKLIST

As a result of your studies of this chapter you should now be able to

1. Differentiate between statically determinate and statically indeterminate beams.

2. Determine the reactions of statically indeterminate beams by the superposition method.

3. Determine the reaction of statically indeterminate beams by the moment-area method.

4. Apply the three-moment equation to determine the reactions of continuous beams.

10-4 SUMMARY

Beams which have a greater number of unknown reactions at their supports than the number of applicable equations from statics are *statically indeterminate.* Reactions that can be removed and still leave the beam supported are called *redundant*. The *measure of redundancy* is the number of reactions (including moments at supports) that must be removed to make the beam statically determinate. If only one reaction need be removed the beam is said to be *statically indeterminate in the first degree*, if two, it is said to be *statically indeterminate in the second degree*, and so on.

The easiest method of determining the unknown reactions of a statically indeterminate beam is the *superposition method*, in which we take advantage of the fact that the sum of the deflections *or* slopes at any point of the beam equals the total deflection or slope at that point.

The *moment-area* method, although more difficult than the superposition method, is especially valuable in determining the reactions of continuous beams with three or more supports. Unlike the superposition method, the power of the moment-area method is unlimited. A significant relationship combining the superposition method and the moment-area method is the *three-moment* equation below.

$$M_A L_1 + 2 M_B (L_1 + L_2) + M_C L_2 = -\frac{6 A_1 \bar{x}_1}{L_1} - \frac{6 A_2 \bar{x}_2}{L_2} \qquad (10\text{-}1)$$

PROBLEMS

In the problems that follow the weight of the beam shall be considered to be insignificant unless otherwise specified. Each problem requires that all reactions at the supports be determined including any moments exerted by the supports on the beams.

10-1C The figure shows a brick wall weighing 500 lb/ft supported by a propped cantilever beam 18 ft long. Determine all reactions by the superposition method.

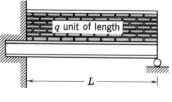

PROBLEM 10-1

STATICALLY INDETERMINATE BEAMS

10-1M   *The figure shows a brick wall weighing 750 kg/m supported by a propped cantilever beam 6 m long. Determine all reactions by the superposition method.*

10-2C   Solve Problem 10-1C by the moment-area method.

10-2M   *Solve Problem 10-1M by the moment-area method.*

10-3C   The figure shows a propped cantilever beam 18 ft long with a concentrated load of 2000 lb at a distance *a* 12 ft from the wall. Determine all reactions by the superposition method.

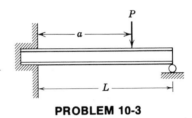

PROBLEM 10-3

10-3M   *The figure shows a propped cantilever beam 6 m long with a concentrated load of 1 000 kg at a distance a 4 m from the wall. Determine all reactions by the superposition method.*

10-4C   Solve Problem 10-3C by the moment-area method.

10-4M   *Solve Problem 10-3M by the moment-area method.*

10-5C   The figure shows a cantilever beam 9 ft long propped at a distance *a* 6 ft from its support with a load *P* of 5000 lb at its end. Determine all reactions by the superposition method.

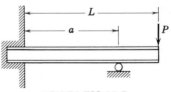

PROBLEM 10-5

10-5M   *The figure shows a cantilever beam 3 m long propped at a distance a 2 m from its support with a load P of 7 500 kg at its end. Determine all reactions by the superposition method.*

10-6C   Solve Problem 10-5C by the moment-area method.

10-6M   *Solve Problem 10-5M by the moment-area method.*

10-7C   Determine the reactions of the beam shown by the superposition method if q is 300 lb/ft and L is 20 ft.

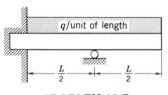

PROBLEM 10-7

10-7M   *Determine the reactions of the beam shown by the superposition method if q is 450 kg/m and L is 6 m.*

10-8C   Solve Problem 10-7C by the moment-area method.

10-8M   *Solve Problem 10-7M by the moment-area method.*

10-9C   Determine the reactions for the beam in the figure by the superposition method if a is 6 ft, b is 2 ft, P is 2000 lb, and q is 200 lb/ft.

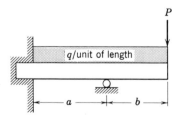

PROBLEM 10-9

10-9M *Determine the reactions for the beam in the figure by the superposition method if a is 3 m, b is 1 m, P is 1 000 kg, and q is 300 kg/m.*

10-10C Solve Problem 10-9C by the moment-area method.

10-10M *Solve Problem 10-9M by the moment-area method.*

10-11C Determine the reactions of the beam in the figure if P_1 is 20,000 lb, P_2 is 10,000 lb, P_3 is 8,000 lb, c is 9 ft, d is 6 ft, and e is 12 ft. Use the superposition method.

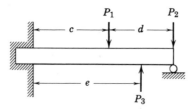

PROBLEM 10-11

10-11M *Determine the reactions of the beam in the figure if P_1 is 9 Mg, P_2 is 4.5 Mg, P_3 is 3.5 Mg, c is 3 m, d is 2 m, and e is 3.5 m. Use the super-position method.*

10-12C Solve Problem 10-11C by the moment-area method.

10-12M *Solve Problem 10-11M by the moment-area method.*

10-13C The figure is a representation of the head casting of a milling machine. The tool arbor is supported by long plain bearings very closely fitted to resist deflection of the arbor. The force P represents the force exerted on the arbor during the milling operation. Determine the reactions at the supports.

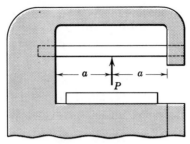

PROBLEM 10-13

10-13M   *Same as Problem 10-13C.*

10-14C   The figure shows a roller mounted on a fixed shaft which is rigidly imbedded at the ends. Determine the reactions at the supports if the roller places a uniform load of 50 lb/ft on the shaft which is 12 in. long.

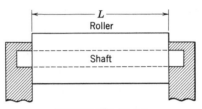

PROBLEM 10-14

10-14M   *The figure shows a roller mounted on a fixed shaft which is rigidly imbedded at the ends. Determine the reactions at the supports if the roller places a uniform load of 75 kg/m on the shaft which is 30 cm long.*

10-15C   A rod is imbedded at the ends as shown in the figure. Determine the reactions at the supports when W is 100 lb, P_1 is 50 lb, P_2 is 200 lb, L_1 is 12 in., L_2 is 16 in., L_3 is 10 in., and L_4 is 8 in.

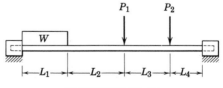

PROBLEM 10-15

10-15M *A rod is imbedded at the ends as shown in the figure. Determine the reactions at the supports when W is 50 kg, P_1 is 25 kg, P_2 is 100 kg, L_1 is 30 cm, L_2 is 40 cm, L_3 is 25 cm, and L_4 is 20 cm.*

10-16C Determine the reactions at the supports of both bars when L_1 is 12 ft, L_2 is 6 ft, P is 5000 lb, and q is 500 lb/ft.

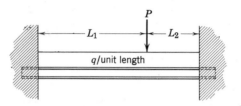

PROBLEM 10-16

10-16M *Determine the reactions at the supports of both bars when L_1 is 4 m, L_2 is 2 m, P is 2500 kg, and q is 750 kg/m.*

10-17C Determine the reactions at the supports of the beam shown in the figure if L is 2 ft, a is 1 ft, q is 50 lb/ft, and W is 50 lb.

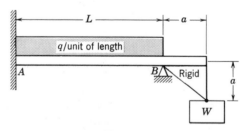

PROBLEM 10-17

10-17M *Determine the reactions at the supports of the beam shown in the figure if L is 60 cm, a is 30 cm, q is 25 kg/m, and W is 25 kg.*

10-18C Determine the reactions at the supports of the horizontal beam in the figure if W is 600 lb, A is 4 ft, B is 3 ft, and c is 1 ft. The weld permits no pivoting at the support.

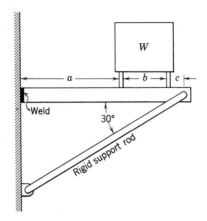

PROBLEM 10-18

10-18M *Determine the reactions at the supports of the horizontal beam in the figure if W is 275 kg, A is 1.2 m, b is 1 m, and c is 0.3 m. The weld permits no pivoting at the support.*

10-19C The figure shows a cantilever beam of a marquee which is supported both by its imbedment in the wall and the rigid cable. Determine the reactions at the supports of the beam if L is 20 ft, the design load on the beam is 100 lb/ft, and the marquee extends one foot beyond the cable attachment.

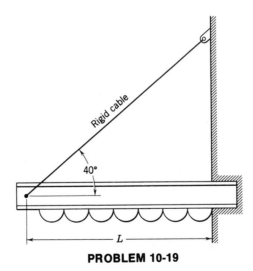

PROBLEM 10-19

10-19M *The figure shows a cantilever beam of a marquee which is supported both by its imbedment in the wall and the rigid cable. Determine the reactions at the supports of the beam if L is 6 m, the design load on the beam is 1 460 N/m, and the marquee extends 30 cm beyond the cable attachment.*

10-20C Determine all reactions of both beams in the figure if *P* is 1000 lb and *L* is 5 ft. Both beams are of the same material and cross section.

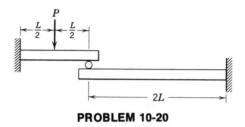

PROBLEM 10-20

10-20M *Determine all reactions of both beams in the figure if P is 450 kg and L is 1.5 m. Both beams are of the same material and cross section.*

10-21C Determine all reactions of the beam in the figure if *P* is 5000 lb, *a* is 10 ft, *b* is 6.5 ft, L_2 is 13 ft, and *q* is 200 lb/ft.

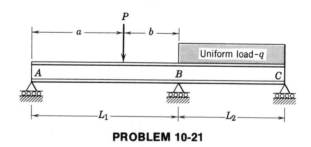

PROBLEM 10-21

10-21M *Determine all reactions of the beam in the figure if P is 2 250 kg, a is 3 m, b is 2 m, L_2 is 4 m, and q is 300 kg/m.*

10-22C Determine all reactions of the beam in the figure if $P_1 = P_2 = 8000$ lb and $L = 30$ ft.

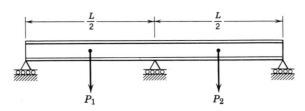

PROBLEM 10-22

10-22M   *Determine all reactions of the beam in the figure if $P_1 = P_2 = 3\,500\,kg$ and $L = 9\,m$.*

10-23C   Determine all reactions of the beam in Problem 10-22C if P_1 is 5500 lb, P_2 is 11,000 lb, and L is 40 ft.

10-23M   *Determine all reactions of the beam in Problem 10-22M if P_1 is $2\,500\,kg$, P_2 is $5\,000\,kg$, and L is $12\,m$.*

10-24C   Determine all reactions on the beam in the figure if a is 10 ft, b is 8 ft, and q is 200 lb/ft.

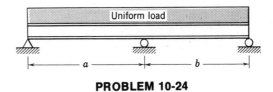

PROBLEM 10-24

10-24M   *Determine all reactions on the beam in the figure if a is $3\,m$, b is $2.5\,m$, and q is $300\,kg/m$.*

10-25C   Determine all reactions of both beams in the figure if q is 275 lb/ft and L is 10 ft.

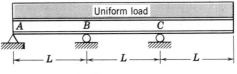

PROBLEM 10-25

277

STATICALLY INDETERMINATE BEAMS

10-25M *Determine all reactions on the beam in the figure if q is 400 kg/m and L is 3 m.*

10-26C Determine all reactions on the beam in the figure if a is 5 ft, b is 10 ft, c is 15 ft, L is 10 ft, P_1 is 5,000 lb, P_2 is 10,000 lb, and P_3 is 5,000 lb.

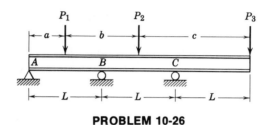

PROBLEM 10-26

10-26M *Determine all reactions on the beam in the figure if a is 1.5 m, b is 3 m, c is 4.5 m, L is 3 m, P_1 is 20 kN, P_2 is 40 kN, and P_3 is 20 kN.*

10-27C Determine all reactions on the beam in the figure if a is 8 ft, b is 10 ft, c is 6 ft, and P is 3000 lb.

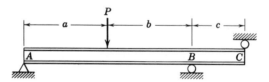

PROBLEM 10-27

10-27M *Determine all reactions on the beam in the figure if a is 3 m, b is 4 m, c is 2 m and P is 1 000 kg.*

10-28C Determine all reactions on the beam in the figure if a is 10 ft, b is 8 ft, P_1 is 10,000 lb, P_2 is 5000 lb, and q is 275 lb/ft.

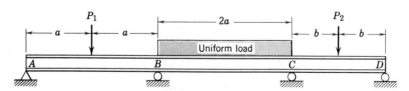

PROBLEM 10-28

278

10-28M *Determine all reactions on the beam in the figure if a is 3 m, b is 2.4 m, P_1 is 4 500 kg, P_2 is 2 250 kg, and q is 400 kg/m.*

10-29C Determine the moments and reactions at the supports of the beam in the figure if L is 20 ft and q is 500 lb/ft.

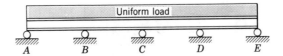

Uniform load

A   B   C   D   E

PROBLEM 10-29

10-29M *Determine the moments and reactions at the supports of the beam in the figure if L is 5 m and q is 700 kg/m.*

CHAPTER
OBJECTIVES

The object of studying this chapter is to learn how to

1. Determine the maximum stresses acting on the internal particles of the material of structural and machine members.

2. How to use Mohr's circle diagram to determine the maximum stresses in structural and machine members carrying combinations of stresses parallel and perpendicular to their axis and shearing stresses due to torsion.

3. Determine the maximum stresses in members exposed to simultaneous axial and bending stresses.

CHAPTER————————————11
COMBINED STRESSES

Although in the majority of cases members of structures and machines sustain loads in one direction only, there are many cases where they are simultaneously subjected to loads of more than one type. For instance, in structures, columns are frequently subjected to bending or eccentric loads at the same time they perform their normal duty of supporting the axial load. In machines members are often loaded axially at the same time they are transmitting a torque; for example, a bolt being wrung tight is withstanding both torsion and tension.

11-1 COMBINED STRESS—GENERAL CASE

To analyze the effect of forces acting in two perpendicular directions combined with torsion on a structure or machine member, we will extract a small rectangular particle shown in Figure 11-1 from the member. The particle is held in equilibrium by the stresses f_x induced by a tensile force acting in the direction of the x axis, f_y induced by a tensile force acting in the direction of the y axis, and the shearing stress f_{vxy}, induced by a turning moment (torsion) on the member. In the analysis that follows f_x and f_y will be considered positive when they represent tension and negative for compression. The shear stress f_{vxy} will be considered to be positive as shown in Figure 11-1 and negative in the opposite direction. It can be proven that the shearing stresses on mutually perpendicular surfaces of an element are equal.

The dashed line mn in Figure 11-1(a) represents any plane perpendicular to

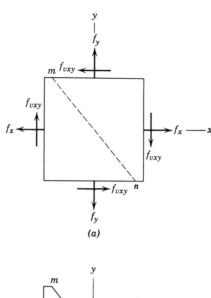

(a)

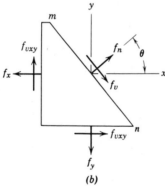

(b)

FIGURE 11-1

the plane of the x and y axes. In Figure 11-1(b) that portion of the particle to the right of mn has been removed. In order for the part to now be held in static equilibrium the shear stress f_v, parallel to the surface, and the tensile stress f_n, perpendicular to the surface, are required. By summing stresses in the f_n direction and taking into consideration that the stress varies with the change in area, we find

$$f_n = \frac{f_x + f_y}{2} + \frac{f_x - f_y}{2} \cos 2\theta - f_{vxy} \sin 2\theta \qquad (11\text{-}1)$$

Similarly, by summing in the direction of f_v we find

$$f_v = \frac{f_x - f_y}{2}\sin 2\theta + f_{vxy}\cos 2\theta \tag{11-2}$$

Further analysis of the above equations determines the maximum normal stress to be

$$f_{n_{max}} = \frac{f_x + f_y}{2} \pm \sqrt{\left(\frac{f_x - f_y}{2}\right)^2 + f_{vxy}^2} \tag{11-3}$$

whichever is larger.

The lesser value of Equation 11-3 is the minimum normal stress $f_{n_{min}}$. The maximum shear stress $f_{v_{max}}$ is

$$f_{v_{max}} = \pm\sqrt{\left(\frac{f_x - f_y}{2}\right)^2 + f_{vxy}^2} \tag{11-4}$$

It can be seen that these equations are rather mathematically tedious. To avoid this we can use the graphical method described in the following section.

11-2 MOHR'S CIRCLE DIAGRAM

Equations 11-1 through 11-4 can be represented graphically by Mohr's circle diagram which was developed for this purpose in 1892 by the German engineer, Otto Mohr. Figure 11-2 shows the format of the diagram. It is drawn as follows.

1. Establish the horizontal f_n and vertical f_v axes and select a convenient scale for plotting. The scale must be the same for both axes. Distances up and to the right of O are positive. Negative values are plotted down and to the left of O.
2. Point B is located on the f_n axis at a distance equal to the stress in the x direction f_x from O. The stress in the y direction is similarly plotted, thus locating point A.
3. From point B measure vertically upward a distance equal to the shear stress f_{vxy}, locating point C. In like manner measure the same distance downward from point A, locating D.
4. By drawing a straight line from C to D locate the center of the circle O', then draw the circle.

The maximum normal stress is found at point F where the circle intersects the f_n axis and the maximum shear stress is found at point G, 90° counterclockwise from F.

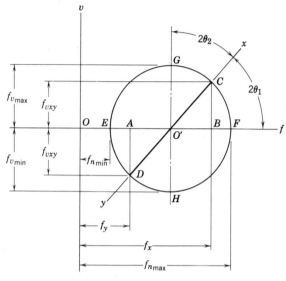

FIGURE 11-2

Example

In the design of a shaft it is found that the stress in the y direction f_y is 1000 units in tension, that in the x direction f_x is 4000 units in tension, and the shearing stress due to torsion is 2000 units. Draw a Mohr's circle diagram to determine the maximum stresses developed in the shaft.

Solution

STEP 1.

Establish the f_v and f_n axes as shown in Figure 11-3 and locate points A and B at distances 1000 and 2000 units, respectively, from point O precisely to scale.

STEP 2.

Plot the shear stress vertically upward from B and downward from A, thereby locating points C and D.

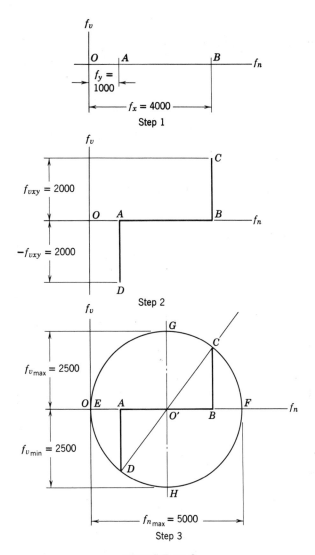

FIGURE 11-3

COMBINED STRESSES

STEP 3.

Draw the line DC thus locating point O' and draw the circle with O' as its center.

By measuring the distance from O to F we find the maximum principle stress $f_{n_{max}}$ to be 5000 units and the maximum shear stress $f_{v_{max}}$ to be 2500 units.

Machine shafting in many situations is exposed to combinations of axial and torsional loads.

Example

The power screws of a tensile test machine are loaded during a tensile test by an axial load of 5000 units compression and a torsional shearing stress of 3000 units. Determine the maximum stresses.

Solution

STEP 1.

In this case the stress in the x direction is compressive. Therefore to locate point B in Figure 11-4 we must measure 5000 units in the negative (compressive) direction from point O. Since the load in the y direction is zero, A is at the origin.

STEP 2.

Points C and D are located by measuring 3000 units upward and downward, respectively, from points B and A. Draw the line CD locating O' and draw the circle with O' as its center.

Measuring the diagram we find the maximum normal stress to be 6400 units compression (to the left of O) and the maximum shear stress to be 3900 units.

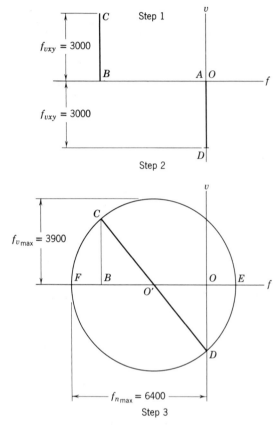

FIGURE 11-4

11-3 COMBINED AXIAL AND BENDING LOADS

The combination of axial and bending loads is very common in both mechanisms and structures. This is best illustrated by the most common example—the ladder. The maximum stress developed by the combination of axial and bending loads is the sum of the bending stress and the direct axial stress at the critical point, the critical point being the point where the highest total stress is developed. In a majority of cases the critical point is apparent. It will generally be at a point where the total shear in bending is zero—where the bending moment is greatest. In the case of the ladder the maximum

287

bending moment will be when the load is at its center. Above the load there would be negligible axial stress, below the load it would be constant down to the ground and equal to the component of the weight on the ladder parallel to the ladder. The following example will demonstrate the technique for solving all such problems.

Example

customary

The ladder in Figure 11-5 is 14 ft long and is made of 1 in. × 4 in. white pine. Determine the maximum stress developed when $a = 5$ ft and $P = 280$ lb.

metric

The ladder in Figure 11-5 is 4 m long and is made of 3 cm × 12 cm material. Determine the maximum stress developed when a is 1 m and P is 100 kg.

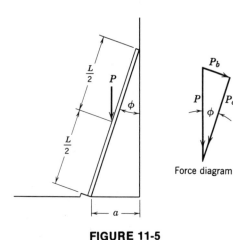

Force diagram

FIGURE 11-5

Solution

customary

It is apparent that the maximum stress will be developed immediately below the point of application of the load. The first step will be to resolve the 280 lb load P

metric

It is apparent that the maximum stress will be developed immediately below the point of application of the load. The first step will be to resolve the into components P_b perpendicular to the ladder and P_a parallel to the ladder. Referring to the force diagram in the figure

$$\sin \phi = \frac{P_b}{P} = \frac{5 \text{ ft}}{14 \text{ ft}} = 0.357,$$

$$P_b = P \sin \phi$$
$$= 280 \text{ lb } (0.357) = 100 \text{ lb}$$

$$\cos \phi = \frac{P_a}{P} = \frac{\sqrt{14^2 - 5^2}}{14} = 0.934,$$

$$P_a = P \cos \phi$$
$$= 280 \text{ lb } (0.934) = 262 \text{ lb}$$

The maximum stress is the axial stress plus the maximum bending stress.

$$f_{max} = \frac{P_a}{A} + \frac{M}{S} \qquad (11\text{-}5)$$

where M is the maximum bending moment, A is the cross-sectional area of the ladder sides, and S is the section modulus of the ladder sides.

The ladder is a simple beam with a concentrated load P_b at its center so

$$M = \frac{P_b L}{2} = \frac{100 \text{ lb } (14 \text{ ft})}{2}$$
$$= 700 \text{ lb-ft} = 8400 \text{ lb-in.}$$

For the rectangular cross section of the ladder sides

$$S = 2\frac{bh^2}{6} = 2\frac{1 \text{ in. } (4 \text{ in.})^2}{6} = 5.33 \text{ in.}^3$$

and

$$A = 2bh = 2 \, (1 \text{ in.})(4 \text{ in.}) = 8 \text{ in.}^2$$

Substituting these values in the above maximum stress equation gives us

$$f_{max} = \frac{262 \text{ lb}}{8 \text{ in.}^2} + \frac{8400 \text{ lb-in.}}{5.33 \text{ in.}^3} = 1610 \text{ psi}$$

100 kg load into the force components P_b perpendicular to the ladder and P_a parallel to the ladder.

$$P = 100 \text{ kg } (9.81 \text{ N/kg}) = 981 \text{ N}$$

Referring to the force diagram in the figure

$$\sin \phi = \frac{P_b}{P} = \frac{1 \text{ m}}{4 \text{ m}}, \qquad P_b = \frac{P}{4} = \frac{981 \text{ N}}{4}$$
$$= 245 \text{ N}$$

$$\cos \phi = \frac{P_a}{P} = \frac{\sqrt{16^2 - 1^2}}{4} = 0.968,$$

$$P_a = (0.968)(981 \text{ N}) = 950 \text{ N}$$

The maximum stress being the axial stress plus the bending stress

$$f_{max} = \frac{P_a}{A} + \frac{M}{S}$$

where M is the maximum bending moment, A is the cross-sectional area of the ladder sides, and S is the section modulus of the ladder sides.

For a simple beam with a concentrated load at its center,

$$M = \frac{P_b L}{2} = \frac{245 \text{ N}(4 \text{ m})}{2} = 490 \text{ N·m}$$

For the rectangular cross section of the ladder sides

$$S = 2\frac{bd^2}{6} = 2\frac{0.03 \text{ m } (0.12 \text{ m})^2}{6}$$
$$= 1.44 \times 10^{-4} \text{ m}^3$$

Also

$$A = 2bd = 2(0.03 \text{ m})(0.12 \text{ m})$$
$$= 7.2 \times 10^{-3} \text{ m}^2$$

Substituting these values in the above maximum stress equation gives us

COMBINED STRESSES

As indicated by the above equation the maximum stress occurs where the axial and bending stresses are both in tension or both in compression. Since the axial stress in this case is compressive the 1610 psi must be a compressive stress. It occurs on the side of the ladder toward the load.

$$f_{max} = \frac{950\,N}{7.2 \times 10^{-3}\,m^2} + \frac{490\,N{\cdot}m}{1.44 \times 10^{-4}\,m^3}$$
$$= 3.53\,MPa$$

As indicated by the above equation the maximum stress occurs where the axial and bending stresses are both in tension or are both in compression. Since the axial stress in this case is compressive the 3.53 MPa must be a compressive stress. It occurs on the side of the ladder toward the load.

11-4 ACCOMPLISHMENT CHECKLIST

On the basis of your study of this chapter you should know how to

1. Determine the maximum normal and shear stress in members acted on by loads acting at right angles in either tension or compression.

2. Determine the maximum normal and shear stress in members of structures or machines acted on by combinations of axial and torsional loads.

3. Determine the maximum normal tensile and compressive stress in structural or machine members acted on by combined axial and bending stresses.

11-5 SUMMARY

If a particle is removed from a body under stress, the normal stress and shear stress acting on any plane within the particle and perpendicular to the plane of the x and y axes of the particle are given by the following equations where the angle θ is the angle between a normal to the plane and the x axis as shown in Figure 11-1.

$$f_n = \frac{f_x + f_y}{2} + \frac{f_x - f_y}{2}\cos 2\theta - f_{vxy}\sin 2\theta \qquad (11\text{-}1)$$

$$f_v = \frac{f_x - f_y}{2}\sin 2\theta + f_{vxy}\cos 2\theta \qquad (11\text{-}2)$$

The maximum f_v and f_n stresses are determined by

$$f_{n_{max}} = \frac{f_x + f_y}{2} \pm \sqrt{\left(\frac{f_x - f_y}{2}\right)^2 + f_{vxy}^2}$$ (11-3)

whichever is larger.

$$f_{v_{max}} = \pm \sqrt{\left(\frac{f_x - f_y}{2}\right)^2 + f_{vxy}^2}$$ (11-4)

The foregoing equations can be graphically represented by the circle diagram developed by Otto Mohr. Combined axial and bending stresses may be determined by adding the maximum bending stress to the axial stress giving

$$f_{max} = \frac{P_a}{A} + \frac{M}{S}$$ (11-5)

where P is the component of the load parallel to the axis of the member, A is the cross-sectional area of the member, S is its section modulus, and M is the maximum bending moment.

PROBLEMS

In the following problems mathematically determine the maximum normal and shear stresses.

	f_x	f_y	$f_{v_{xy}}$
11-1C	0	20,000 psi	0
11-1M	0	140 MPa	0
11-2C	10,000 psi	10,000 psi	10,000 psi
11-2M	70 MPa	70 MPa	70 MPa
11-3C	0	0	10,000 psi
11-3M	0	0	70 MPa

COMBINED STRESSES

11-4C	10,000 psi	20,000 psi	5,000 psi
11-4M	*70 MPa*	*140 MPa*	*35 MPa*
11-5C	10,000 psi	20,000 psi	0
11-5M	*70 MPa*	*140 MPa*	*0*

11-6C   Solve Problem 11-1C using Mohr's circle diagram.

11-6M   *Solve Problem 11-1M using Mohr's circle diagram.*

11-7C   Solve Problem 11-2C using Mohr's circle diagram.

11-7M   *Solve Problem 11-2M using Mohr's circle diagram.*

11-8C   Solve Problem 11-3C using Mohr's circle diagram.

11-8M   *Solve Problem 11-3M using Mohr's circle diagram.*

11-9C   Solve Problem 11-4C using Mohr's circle diagram.

11-9M   *Solve Problem 11-4M using Mohr's circle diagram.*

11-10C   Solve Problem 11-5C using Mohr's circle diagram.

11-10M   *Solve Problem 11-5M using Mohr's circle diagram.*

11-11C   During the compression test of a concrete cylinder the power screw of the test machine undergoes a tensile stress of 4000 psi and a torsional stress of 1000 psi. Determine the maximum normal and shear stress in the material.

11-11M *During the compression test of a concrete cylinder the power screw of the test machine undergoes a tensile stress of 28 MPa and a torsional stress of 7 MPa. Determine the maximum normal and shear stress in the material.*

11-12C The supporting column of a rotating swing carnival ride is designed to support a maximum total load 10,000 lb while withstanding a maximum driving torque of 20,000 lb-in. Determine the maximum normal and shear stress developed in the column which is a solid steel shaft 5 in. in diameter.

11-12M *The supporting column of a rotating swing carnival ride is designed to support a maximum total load of 5000 kg while withstanding a maximum driving torque of 2000 N·m. Determine the maximum normal and shear stress developed at the design loads on the column which is a solid steel shaft 125 mm in diameter.*

11-13C Determine the minimum diameter the shaft in Problem 11-12C could be if the allowable normal stress for the material is 80,000 psi and the maximum shear stress is 60,000 psi. Use a safety factor of 2.5.

11-13M *Determine the minimum diameter the shaft in Problem 11-12M could have if the allowable maximum normal stress for the material is 550 MPa and the maximum shear stress is 415 MPa. Use a safety factor of 2.5.*

11-14C To drive a 1 in. drill through mild steel with a feed of 0.013 in./rev requires a thrust of about 1415 lb and a torque of about 650 in.-lb. Determine the maximum stresses developed in the shank (shank diameter is the same as the drill diameter).

11-14M *To drive a 25 mm drill through mild steel with a feed of 0.33 mm/rev requires a thrust of about 6295 N and a torque of about 74 N·m. Determine the maximum stresses developed in the shank (the shank diameter is the same as the diameter of the drill itself).*

11-15C Determine the maximum stress developed in the beam when P is 8000 lb, L is 15 ft, the cross-sectional area is 19.1 in.^{2} and the section modulus of the beam section is 88 in.^{4}

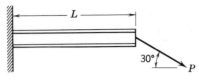

PROBLEM 11-15

11-15M *Determine the maximum stress developed in the beam when P is 4000 kg, L is 5 m, the cross-sectional area is 1.23×10^{-2} m, and the section modulus of the beam section is 1.44×10^{-3} m^3.*

11-16C Determine the maximum stress in the beam in the figure when F is 2000 lb, L is 20 ft, x is 10 ft, and the cross section of the beam is 4 in. wide $\times$ 8 in. high.

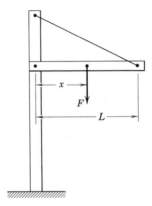

PROBLEM 11-16

11-16M *Determine the maximum stress in the beam in the figure when F is 1000 kg, L is 6 m, x is 3 m, and the cross section of the beam is 10 cm wide $\times$ 20 cm high.*

11-17C Same as Problem 11-16C except that x is 15 ft.

11-17M *Same as 11-16M except that x is 5 m.*

11-18C Determine the maximum stress in the beam AB when W is 1000 lb, a is 8 ft, b is 12 ft, the cross-sectional area of the beam is 8.54 $in.^2$ and the section modulus of AB is 30.8 $in.^4$

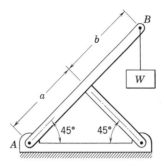

PROBLEM 11-18

11-18M *Determine the maximum stress in the beam AB when W is 450 kg, a is 2.5 m, b is 3.5 m, the cross-sectional area of the beam is $5.51 \times 10^{-3} m^2$, and the section modulus of AB is $5.05 \times 10^{-4} m^3$.*

11-19C Determine the maximum stresses developed in the capstan power shaft when P is 32,000 lb, h is 2 ft, the torque being transmitted by the shaft is 630,000 lb-in. The power shaft is hollow with an O.D. of 10 in. and an I.D. of 9 in.

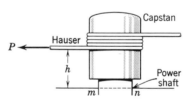

PROBLEM 11-19

11-19M *Determine the maximum stresses developed in the capstan power shaft when P is 140 kN, h is 0.7 m, the torque being transmitted by the shaft is 70 kN·m. The power shaft is hollow with an O.D of 25 cm and an I.D. of 20 cm.*

11-20C The helical gear in the figure transmits a torque T of 3600 lb-in. Due to the angle of the gear teeth, the mating gear (not shown) exerts a thrust A parallel to the axis of the 1 in. diameter shaft of 350 lb at the same time it exerts a force B of 640 lb perpendicular to the axis of the shaft. The distance between the centers of the bearings which support the shaft is 12 in., with the gear being mounted halfway between the bearings. Determine the maximum shear and normal stresses in the shaft.

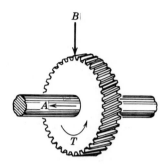

PROBLEM 11-20

11-20M *The helical gear in the figure transmits a torque T of 400 N·m. Due to the angle of the gear teeth, the mating gear (not shown) exerts a thrust A parallel to the axis of the 40 mm diameter shaft of 1500 N at the same time it exerts a force B of 2800 N perpendicular to the axis of the shaft. The distance between the centers of the bearings which support the shaft is 0.300 m, with the gear being mounted halfway between the bearings. Determine the maximum shear and normal stresses in the shaft.*

CHAPTER
OBJECTIVES

The purpose of studying this chapter is to learn

1. The nature of columns.

2. The nature of instability.

3. The classification of columns.

4. About Euler's column theory and its application.

5. How to design columns using empirical formulas.

CHAPTER————————————12
COLUMNS

A *column* is a structural or machine member which carries a compressive load in the direction of its axis. Unlike beams which usually signal their oncoming failure by visible changes such as excessive bending, columns usually fail instantaneously and catastrophically. On top of that the safe load for a column is much more difficult to determine because imperfections in the material and variations of the dimensions of a column have significant effects on their load-bearing abilities. A *perfect column* would be absolutely straight, perfectly uniform in its cross section, loaded precisely at the centroid of its cross section, and its material would be perfectly uniform throughout with no flaws or residual stresses. Obviously, no perfect column exists. Research in the design of columns is continuous and is producing a continual upgrading of design knowledge. The information in this chapter is very basic and necessarily limited.

12-1 CHARACTERISTICS OF COLUMN FAILURES

Figure 12-1(a) shows a *post*. Post fail without buckling due to excessive axial loads. Usually the failure is by shearing as in Figure 12-1(b), although it could be simply by crushing. For our purposes we will define a column as a structural or machine member which will collapse by buckling if exposed to an excessive axial compressive load. Figure 12-1(c) shows a column safely loaded, whereas Figure 12-1(d) shows an overloaded column. When such a column develops anything more than a very slight bend it will collapse

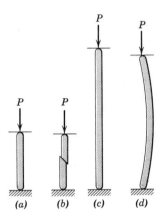

FIGURE 12-1

instantaneously. A failure of this type is one of instability—after it passes a certain point of bending there is no possibility of recovery. The idea of stability versus instability is shown in Figure 12-2. In the left-hand figure are two balls in a trough. They have gravitated to the low point of the trough and have taken positions in equilibrium, demonstrating stability. If the trough is inverted as in the right-hand figure, balls placed on it will roll off never to attain equilibrium within the realm of the system, thus demonstrating instability.

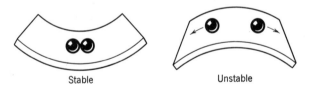

Stable   Unstable

FIGURE 12-2

12-2 EULER'S COLUMN THEORY

Well over 200 years ago the famous Swiss mathematician, Leonhard Euler, developed the following equation for the critical load for a perfect column, the *critical load* being that load above which the beam would be unstable and subject to buckling.

$$P_c = \frac{\pi^2 EI}{L^2} \tag{12-1}$$

To obtain a more easily usable form of this equation we can substitute $r^2 A$ for I where the r is the least radius of gyration of the column and A is its cross-sectional area. This gives us

$$P_c = \frac{\pi^2 E r^2 A}{L^2}$$

Now divide both sides by A

$$\frac{P_c}{A} = \frac{\pi^2 E r^2}{L^2} \quad \text{or} \quad f_c = \frac{\pi^2 E}{(L/r)^2} \tag{12-2}$$

where f_c is the *critical stress* (the stress at load P_c) and L/r is called the *slenderness ratio*. As previously noted this equation is based on a perfect column. In addition it is limited to a column of which the ends are free to pivot. To account for different end mounting conditions an *effective length*, KL, is used where K is the *effective length factor* as given in Table 12.1.

TABLE 12-1

	(a)	(b)	(c)	(d)	(e)	(f)
Buckled shape of column is shown by dashed line						
Theoretical K value	0.5	0.7	1.0	1.0	2.0	2.0
Recommended K value when ideal conditions are approximated	0.65	0.80	1.2	1.0	2.10	2.0
End condition code		Rotation fixed, Translation fixed				
		Rotation free, Translation fixed				
		Rotation fixed, Translation free				
		Rotation free, Translation free				

Source. GUIDE TO STABILITY DESIGN CRITERIA FOR METAL STRUCTURES, 3rd ed., Bruce G. Johnston, ed., Copyright © 1966, John Wiley & Sons, New York. Reprinted by permission.

This effective length is then inserted in Equation 12-2 as follows

$$f_c = \frac{\pi^2 E}{(KL/r)^2} \qquad (12\text{-}2a)$$

Up to now we have been dealing with perfect columns. As made, columns are not perfect—they do contain residual stresses due to the rolling and cooling processes of their manufacture—they may contain residual stresses due to fabrication operations such as welding subsequent to the manufacture—they are not perfectly straight—their material may not be the same throughout or have exactly the same properties in every direction—the load probably is not precisely applied at their centroidal axes. To provide for these deviations from perfection a safety factor must be applied. The American Institute of Steel Construction recommends* that Equation 12-3 be divided by a safety factor of 23/12. Inserting the safety factor in Equation 12-2a gives us the practical equation for the allowable stress for long columns.

$$F_a = \frac{12\,\pi^2 E}{23\,(KL/r)^2} \qquad (12\text{-}3)$$

A *long column* is one for which

$$\frac{KL}{r} \geq \sqrt{\frac{2\pi^2 E}{f_y}} \qquad (12\text{-}4)$$

The term on the right side of the above equation is C_c, the column slenderness ratio dividing elastic and inelastic buckling. Equation 12-3 can be used only for long columns. The following examples demonstrate the use of Equations 12-3 and 12-4 in the designing of long columns.

Example

customary

Select a 24 ft long column which is pinned at each end and must support a 64 ksi load from the wide-flanged shapes in Appendix 2. The steel is to be A36 with a yield strength of 36 kip and an E of 29,000,000 psi.

metric

Select a 7.5 m long column which is pinned at each end and must support a 30 Mg load from the wide-flanged shapes in Appendix 2. The steel is to be A36 with a yield strength of 248 MPa and an E of 200 GPa.

* American Institute of Steel Construction, "Specification for the Design, Fabrication and Erection of Structural Steel for Buildings," 1969; Supplement No. 1, 1970, and Supplement No. 2, 1971.

Solution

customary

The procedure is basically trial and error. The first step is to determine the cross-sectional area required to resist direct compression.

$$A = \frac{P}{F_a} = \frac{64 \, \text{kip}}{36 \, \text{ksi}} = 1.78 \, \text{in.}^2$$

To resist buckling a much larger area will be required. A relatively square column will probably be the most economical, so we will select a W6X15.5 section having an r of 1.46 in. and an A of 4.56 in.^{2} from those in Appendix 2.

For this section

$$\frac{KL}{r} = \frac{(1)(288 \, \text{in.})}{1.46 \, \text{in.}} = 197$$

C_c for A36 steel is

$$\sqrt{\frac{2\pi^2 E}{f_y}} = \sqrt{\frac{2\pi^2(29 \times 10^6 \, \text{psi})}{36,000 \, \text{psi}}} = 126$$

Since our KL/r is larger than C_c we are dealing with a long column and can use the Euler equation (12-3) to determine the allowable stress.

$$F_a = \frac{12\pi^2 E}{23\left(\frac{KL}{r}\right)^2} = \frac{12\pi^2(29 \times 10^6 \, \text{psi})}{23(197)^2}$$
$$= 3850$$

$$P_a = F_a A = 3850 \, \text{psi}(4.56 \, \text{in.}^2)$$
$$= 17{,}500 \, \text{lb}$$

The allowable load is less than half the actual load. We must select a larger section. For our next try we will select a W8X31 having an r of 2.01 in. and an A of 9.12 in.^{2}.

metric

The procedure is basically trial and error. The first step is to determine the cross-sectional area required to resist direct compression.

$$A = \frac{P}{F_a} = \frac{30 \, \text{Mg} \, (9.81 \, \text{N/kg})}{248 \, \text{MPa}}$$
$$= 1.19 \times 10^{-3} \, m^2$$

To resist buckling a much larger area will be required. A relatively square column will probably be the most economical so we will select a $150 \times 150 - 17.2$ section having an r of 3.75 cm and an A of 40.14 cm^2.

For this section

$$\frac{KL}{r} = \frac{1 \, (7.5 \, m)}{0.0375 \, m} = 200$$

C_c for A36 steel is

$$\sqrt{\frac{2\pi^2 E}{f_y}} = \sqrt{\frac{2\pi^2 \, (200 \, GPa)}{248 \, MPa}} = 126$$

Since our KL/r is larger than C_c we are dealing with a long column and can use the Euler equation (12-3) to determine the allowable stress.

$$F_a = \frac{12\pi^2 E}{23 \, (KL/r)^2} = \frac{12\pi^2 \, (200 \, GPa)}{23 \, (200)^2}$$
$$= 12.7 \, MPa$$

$$P_a = F_a A = 12.7 \, MPa \, (0.4014 \, m^2)$$
$$= 5.10 \, MN = 0.520 \, Mg$$

This is far lower than the actual load. We must select a larger section. For our next try we will select a $200 \times 200 - 56.2$ wide flange section having an r of 5.02 cm and an A of 63.53 cm^2.

For this section

$$KL/r = 1(228 \text{ in.})/2.01 \text{ in.} = 143$$

This being greater than 126, we still have a long column.

$$F_a = \frac{12\pi^2 E}{23\left(\frac{KL}{r}\right)^2} = \frac{12\pi^2(29 \times 10^6 \text{ psi})}{23(142)^2}$$

$$= 7300 \text{ psi}$$

$$P_a = F_a A = 7300 \text{ psi}(9.12 \text{ in.}^2) = 66.6 \text{ kip}$$

This is the best to be found in Appendix 2. It should be realized that a closer value might be found in a complete table furnished by the steel producer.

For this section

$$\frac{KL}{r} = \frac{1\,(7.5\,m)}{0.0\,502\,m} = 149$$

This being greater than 126, we still have a long column.

$$F_a = \frac{12\pi^2\,(200\,GPa)}{23(149)^2} = 46.4\,MPa$$

$$P_a = F_a A = 46.4\,MPa\,(0.6\,353\,m^2)$$
$$= 295\,kN = 30\,Mg$$

The $200 \times 200 - 56.2\,kg/m$ column will be just right for the job.

12-3 DESIGN OF COLUMNS USING EMPIRICAL FORMULAS

In the case of columns for which

$$\frac{KL}{r} \leqslant \sqrt{\frac{2\pi^2 E}{f_y}} \tag{12-5}$$

the Euler column formula of the previous section is not usable, and empirical equations must be used. For building columns the AISC recommends the use of an empirical equation divided by a factor of safety. This equation, usable for values of KL/r from 0 to 200, is

$$f_c = f_y - \frac{f_y^2}{4\pi^2 E}\left(\frac{KL}{r}\right)^2 \div \text{F.S.} \tag{12-6}$$

where the factor of safety F.S. is given by

$$\text{F.S.} = \frac{5}{3} + \frac{3\,(KL/r)}{8C_c} - \frac{(KL/r)^3}{8C_c^3} \tag{12-7}$$

where

$$C_c = \sqrt{\frac{2\pi^2 E}{f_y}} \tag{12-8}$$

The above equations make allowance for the deviations of the column from

304

being a perfect column. The most important of these deviations is that caused by the residual stresses developed by the cooling of the column following the rolling process by which it was produced.

Example

customary

A wide-flange column 15 ft long is needed to support a concentric axial load of 350,000 lb. The steel is A36 with a yield strength of 36,000 psi and an E of 29,000,000 psi. Determine the most economical column from the wide-flange shapes in Appendix 2.

metric

A wide-flange column 4.5 m long is needed to support a concentric axial load of 160 Mg. The steel is A36 with a yield strength of 248 MPa and an E of 200 GPa. Determine the most economical column from the wide-flange shapes in Appendix 2.

Solution

customary

The procedure is trial and error. However, if we approach the problem systematically, we can keep the number of trials needed within reason.

First, if we look at Equation 12-7, we find that the factor of safety F.S. will vary between 1.67 when KL/r is zero, to 1.92 when KL/r is equal to C_c. Then if we assume that the second term on the right side of Equation 12-6 is probably small compared with f_y alone, it would seem that we might begin by letting f_c equal f_y divided by the median F.S., which is

$$\frac{1.67 + 1.92}{2} \quad \text{or} \quad 1.80$$

Let $f_c = f_y/1.80$ which for the A36 steel is

$$\frac{36,000 \text{ psi}}{1.80} = 20,000 \text{ psi.}$$

metric

The procedure is trial and error. However, if we approach the problem systematically, we can keep the number of trials needed within reason.

First, if we look at Equation 12-7, we find that the factor of safety F.S. will vary between 1.67 when KL/r is zero to 1.92 when KL/r equals C_c. Then if we assume that the second term on the right side of Equation 12-6 is probably small compared with the free-standing f_y, it would seem that we might begin by letting f_c equal f_y divided by the median F.S.

Let $f_c = f_y/1.80$ which for the A36 steel is

$$\frac{248 \text{ MPa}}{1.80} = 138 \text{ MPa.}$$

At $f_c = 20,000$ psi the required area would be

$$\frac{350,000 \text{ lb}}{20,000 \text{ psi}} = 17.5 \text{ in.}^2$$

We now look in Appendix 2 for a section, relatively square, that has an area close to 17.5 in.^{2} We find that a W12X65 with an A of 19.1 in.^{2} and a least r of 3.02 fills the bill. Using this

$$\frac{KL}{r} = \frac{1\,(180 \text{ in.})}{3.02 \text{ in.}} = 59.6$$

Knowing that C_c is 126 for A36 steel we can proceed to determine F.S.

$$\text{F.S.} = \frac{5}{3} + \frac{3KL/r}{8C_c} - \frac{(KL/r)^3}{8C_c^3}$$

$$\text{F.S.} = \frac{5}{3} + \frac{3\,(59.6)}{8\,(126)} - \frac{59.6^3}{8\,(126)^3}$$

$$= 1.83$$

Then

$$f_c = \left[36,000 \text{ psi} - \frac{(36,000 \text{ psi})^2}{4\pi^2(29 \times 10^6 \text{ psi})}(59.6)^2 \right]$$

$$\div 1.83 = 17,500 \text{ psi}$$

The load that could be safely carried by the W12X65 member would be

$$P = Af_c = 19.1 \text{ m}^2\,(17,500 \text{ psi})$$

$$= 334,000 \text{ lb}$$

This is close but not close enough. For our next try let us go to the next heavier section, a W12X72 with a least r of 3.04 and an A of 21.2.

$$\frac{KL}{r} = \frac{1\,(180 \text{ in.})}{3.04} = 59.2$$

$$\text{F.S.} = \frac{5}{3} + \frac{3\,(59.2)}{8\,(126)} - \frac{(59.2)^3}{8\,(126)^3}$$

$$= 1.83$$

At $f_c = 138$ MPa the required area would be

$$\frac{160\,000 \text{ kg } (9.81)}{138 \text{ MPa}} = 114 \text{ cm}^2.$$

We now look in Appendix 2 for a section, relatively square, that has an area close to 114 cm^2. We find that a 300X300 − 94.0 with an A of 119.8 cm^2 and a least r of 7.51 cm fills the bill. Using this

$$\frac{KL}{r} = \frac{1\,(4.5 \text{ m})}{0.0751 \text{ m}} = 59.9$$

Knowing that C_c is 126 for A36 steel we can proceed to determine F.S.

$$\text{F.S.} = \frac{5}{3} + \frac{3KL/r}{8C_c} - \frac{(KL/r)^3}{8C_c^3}$$

$$\text{F.S.} = \frac{5}{3} + \frac{3\,(59.9)}{8\,(126)} - \frac{59.9^3}{8\,(126)^3}$$

$$= 1.83$$

Then

$$f_c = \left[248 \text{ MPa} - \frac{(248 \text{ MPa})^2}{4\pi^2\,(200 \text{ GPa})}(59.9)^2 \right]$$

$$\div 1.83 = 120 \text{ MPa}$$

The load that could be safely carried by the 300X300 − 94 member would be

$$P = Af_c \text{ where } A \text{ is } 119.8 \text{ cm}^2$$

$$\times 10^{-4} \text{ m}^2/\text{cm}^2$$

$$= 119.8 \times 10^{-4} \text{ m}^2$$

giving us

$$P = 119.8 \times 10^{-4} \text{ m}^2\,(120 \text{ MPa})$$

$$= 1.44 \text{ MN} \quad \text{or} \quad 147\,000 \text{ kg}$$

This is close but not close enough. For our next try let us go to the next heavier section, a 300X300 − 156 with a least r of 7.57 cm and an A of 134.8 cm^2.

$f_c =$

$$\left[36{,}000\text{ psi} - \frac{(36{,}000\text{ psi})^2}{4\pi^2(29{,}000{,}000\text{ psi})}(59.2)^2\right]$$

$\div 1.83 = 17{,}500\text{ psi}$

$P = Af_c = 21.2\text{ in.}^2(17{,}500\text{ psi})$
$= 371{,}000\text{ lb}$

This is satisfactory. Use the W12X72 section.

$$\frac{KL}{r} = \frac{1(4.5\,m)}{0.0\,757\,m} = 59.4$$

$$F.S. = \frac{5}{3} + \frac{3\,(59.4)}{8\,(126)} - \frac{(59.4)^3}{8\,(126)^3}$$
$= 1.83$

Then

$$f_c = \left[248\,MPa - \frac{(248\,MPa)^2}{4\pi^2(200\,GPa)}(59.4)^2\right]$$

$\div 1.83 = 120\,MPa$

$P = Af_c = 134.8 \times 10^{-4}\,m^2\,(120\,MPa)$
$= 1.62\,MN = 165\,Mg$

The $300X300 - 156$ section with the least r of $7.57\,cm$ is satisfactory.

12-4 ACCOMPLISHMENT CHECKLIST

On the basis of your studies of this chapter you should now understand

1. What constitutes a perfect column.

2. The difference between a column and a post.

3. The difference between stability and instability as it pertains to columns.

4. The nature of column failure.

5. The meaning and significance of the terms *critical load, critical stress, slenderness ratio, effective length,* and *effective slenderness ratio.*

6. How to compensate for different end mountings of columns.

7. How to design a long column by using the Euler column equation.

8. How to use the AISC column formulae to design an intermediate column.

9. How to determine whether a column is a long or intermediate column and whether the Euler or AISC equation should be used in its design.

12-5 SUMMARY

A *perfect column* is absolutely straight, perfectly uniform in cross section and material, loaded precisely at the centroid of its cross section, and is completely free of flaws and residual stresses. A member that fails under a direct axial load without buckling is a *post*. A member that buckles under a direct axial load is a *column*. When a member passes a certain point in bending beyond which there is no possibility of recovery, it is said to be *unstable*.

The load above which a column will become unstable is the *critical load*. The critical load for a perfect column, determined mathematically by Leonhard Euler, is

$$P_c = \frac{\pi^2 EI}{L^2} \tag{12-1}$$

A more useful form of the above equation is

$$f_c = \frac{\pi^2 E}{(L/r)^2} \tag{12-2}$$

The term L/r in the above equation is known as the *slenderness ratio*. To account for different end mounting conditions the L in the slenderness ratio is multiplied by an *effective length factor K*, giving us an *effective length KL*. In practical design the effective length is used in the Euler equation along with a safety factor of 23/12 giving us

$$f_c = \frac{12\pi^2 E}{23\,(KL/r)^2} \tag{12-3}$$

Equation 12-3 is usable only for columns for which

$$\frac{KL}{r} \geq \sqrt{\frac{2\pi^2 E}{f_y}} \tag{12-4}$$

The KL/r term above is the *effective slenderness ratio*. Columns for which Equation 12-4 is true are called *long columns*.

Columns for which

$$\frac{KL}{r} \leq \sqrt{\frac{2\pi^2 E}{f_y}} \tag{12-5}$$

are called *intermediate columns* and empirical equations must be used. The AISC recommends the following empirical equation for columns of buildings:

$$f_c = f_y - \frac{f_y^2}{4\pi^2 E}\left(\frac{KL}{r}\right)^2 \div \text{F.S.}$$

where the factor of safety F.S. is given by

$$\text{F.S.} = \frac{5}{3} + \frac{3\,(KL/r)}{8C_c} - \frac{(KL/r)^3}{8C_c^3} \tag{12-7}$$

where

$$C_c = \sqrt{\frac{2\pi^2 E}{f_y}} \tag{12-8}$$

PROBLEMS

Euler's equation problems

12-1 through 12-5. Determine the lowest slenderness ratio for which Euler's equation (12-3) may be used for each of the following.

	Material	Yield Stress f_y	Modulus of Elasticity E
12-1C	Structural Steel	36,000 psi	29,000,000 psi
12-2C	Machinery Steel	80,000 psi	30,000,000 psi
12-3C	Wrought Aluminum	50,000 psi	10,000,000 psi
12-4C	Aluminum Bronze	62,000 psi	18,000,000 psi
12-5C	Titanium	120,000 psi	15,000,000 psi
12-1M	Structural Steel	248 MPa	200 GPa
12-2M	Machinery Steel	552 MPa	207 GPa
12-3M	Wrought Aluminum	345 MPa	69 GPa
12-4M	Aluminum Bronze	427 MPa	124 GPa
12-5M	Titanium	830 MPa	103 GPa

COLUMNS

12-6C   A 2 in. steel shaft 6 ft long is to be used as a compression member in a machine. Determine the maximum load it can safely carry if the steel has a yield strength of 80,000 psi and an E of 30,000,000 psi. The member can pivot at both ends.

12-6M   *A 50 mm steel shaft 2 m long is to be used as a compression member in a machine. Determine the maximum load it can safely carry if the steel has a yield strength of 552 MPa and an E of 207 GPa. The member can pivot at both ends.*

12-7C   Determine the longest W8X24 column, pinned at both ends, that can safely support a 30 kip axial compressive load. The steel is A36 with $f_y = 36,000$ psi and $E = 29,000,000$ psi.

12-7M   *Determine the longest $300 \times 150 - 36.7$ kg/m column, pinned at both ends, having an f_y of 248 MPa, and an E of 200 GPa that will support a load of 13.5 Mg.*

12-8C   Determine the largest compressive axial load that may safely be placed on an aluminum rod 1/2 in. in diameter and 18 in. long if it is made of 2014-T6 aluminum with a yield stress in compression of 53 ksi and an E of 10,900 ksi. The member is pivoted at both ends.

12-8M   *Determine the largest compressive axial load that may safely be placed on an aluminum rod 12 mm in diameter and 40 cm long if it is made of 2014-T6 aluminum with a yield stress in compression of 365 MPa and an E of 75 GPa. The member is pinned at both ends.*

12-9C   Solve Problem 12-8C for the condition that the member is fixed at one end and free at the other.

12-9M   *Solve Problem 12-8M for the condition that the member is free at one end and fixed at the other.*

12-10C   Solve Problem 12-8C for the condition that the member is fixed at both ends.

12-10M   *Solve Problem 12-8M for the condition that the member is fixed at both ends.*

AISC formula problems

12-11C   A 2 in. diameter aluminum rod 36 in. long with both ends fixed is to be used in compression as a structural member. Determine the maximum load that it can support if its f_y is 19 ksi in compression and E is 10,100 ksi.

12-11M   *A 50 mm diameter aluminum rod 0.90 m long with both ends fixed is to be used in compression as a structural member. Determine the maximum axial load it can support if f_y is 131 MPa and E is 69.6 GPa.*

12-12C   Determine the maximum axial compressive load that an 18 ft W8X24 column, fixed at both ends, can safely support if it is made of A36 steel having an f_y of 36,000 psi and an E of 29,000,000 psi.

12-12M   *Determine the maximum axial compressive load that a 5.5 m $300 \times 150 - 36.7$ wide-flange column, fixed at both ends, can safely support if it is made of A36 steel having an f_y of 248 MPa and an E of 200 GPa.*

12-13C   Determine the largest compressive axial load that may safely be placed on a phosphor bronze rod 3/4 in. in diameter and 18 in. long which is pinned at both ends. The phosphor bronze has an f_y of 65,000 psi and an E of 15,000,000 psi.

12-13M   *Determine the largest compressive axial load that may safely be placed on a phosphor bronze rod 50 mm in diameter and 0.45 m long, which is pinned at both ends. The phosphor bronze has an f_y of 448 MPa and an E of 103 GPa.*

12-14C   Solve Problem 12-13C for the condition that the rod is fixed at one end and pinned at the other.

12-14M   *Solve Problem 12-13M for the condition that the rod is fixed at one end and pinned at the other.*

COLUMNS

12-15C   Solve Problem 12-13C for the condition that the rod is fixed at both ends.

12-15M   *Solve Problem 12-13M for the condition that the rod is fixed at both ends.*

design problems

12-16C   A vertical column 12 ft long, fixed at the bottom and pinned at the top, must support a 400,000 lb load acting in the direction of its longitudinal axis. Determine the most economical wide-flange section in Appendix 2 to serve this purpose if it is made of A36 steel with an f_y of 36,000 psi and an E of 29×10^6 psi.

12-16M   *A vertical column 3.5 m long, fixed at the bottom and pinned at the top, must support a 180,000 kg load acting in the direction of its longitudinal axis. Determine the most economical wide-flange section in Appendix 2 to serve this purpose if it is made of A36 steel having an f_y of 248 MPa and an E of 200 GPa.*

12-17C   Solve Problem 12-16C for the condition that the column is laterally supported to prevent bending in its weaker direction while it is free to bend about the x-x axis of its cross section as illustrated in Appendix 2.

12-17M   *Solve Problem 12-16M for the condition that the column is laterally supported to prevent bending in its weaker direction while it is free to bend about the x-x axis of its cross section as illustrated in Appendix 2.*

12-18C   Determine the minimum diameter permissible for the piston rod in the figure. It is 7 in. long. The piston is 2-1/4 in. in diameter. The piston rod material is SAE 4130 steel with a yield stress of 125,000 psi and an E of 30,000,000 psi. Design to take care of a 300 percent overload.

12-18M   *Determine the minimum diameter permissible for the piston rod in the figure. It is 18 cm long. The piston is 60 mm in diameter. The piston rod material is SAE 4130 steel with a yield stress of 860 MPa and an E of 207 GPa. Design to take care of a 300 percent overload. (200 psi = 1.38 MPa)*

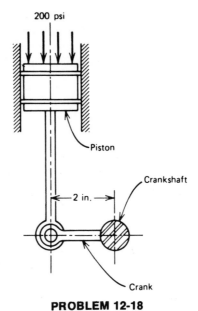

PROBLEM 12-18

12-19C Determine the minimum dimensions for the support rod of the bracket in the figure when P is 200 lb and L is 20 in. The rod is made of 1045 steel with an f_y of 50,000 psi and an E of 30,000,000 psi.

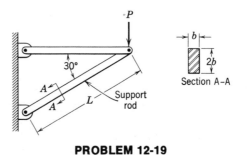

PROBLEM 12-19

12-19M *Determine the minimum dimensions for the support rod of the bracket in the figure when P is 1 000 N and L is 0.5 m. The rod is made of 1045 steel with an f_y of 345 MPa and an E of 207 GPa.*

12-20C Three guy wires support the flag mast on the peak of the building in the figure. There is 300 lb tension in each of the guy wires. The mast is to be an aluminum rod having an f_y in compression of 20 ksi and an E of 10,900 ksi. Determine the minimum permissible diameter of the mast.

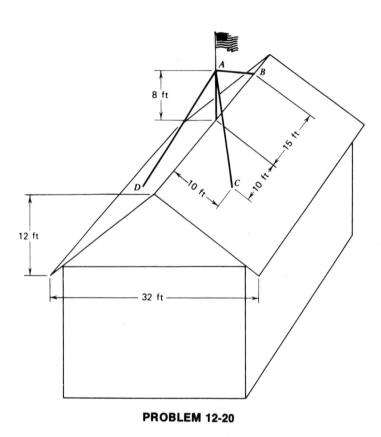

PROBLEM 12-20

12-20M Three guy wires support the flag mast on the peak of the building in the figure. There is 1 500 N tension in each of the guy wires. The mast is to be an aluminum rod having an f_y of 138 MPa and an E of 75 GPa. Determine the minimum permissible diameter for the mast.

12-21C The figure shows the schematic diagram of a support structure. The member *CD* is 20 ft long, *a* is 20 ft, *b* is 30 ft, and *P* is 3000 lb. Determine the most economical *C* section from Appendix 2 for member *CD* if it is made of A36 steel and has an f_y of 36,000 psi and an *E* of 29,000,000 psi.

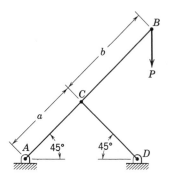

PROBLEM 12-21

12-21M *The figure shows the schematic diagram of a support structure. The member CD is 8.5 m long, a is 8.5 m, b is 13 m, and P is 30 kN. Determine the most economical channel from Appendix 2 for the member CD if it is made of A36 steel with an f_y of 248 MPa and an E of 200 GPa.*

12-22C The figure shows a vertical rod, square in cross section, supporting a horizontal rod carrying the load, *P*, of 500 lb. Determine the required size of the vertical rod to the nearest 1/16 in. when *a* is 4 in., *b* is 6 in., and *L* is 5 in. The rod is steel with $E = 30 \times 10^6$ psi and $f_y = 50,000$ psi.

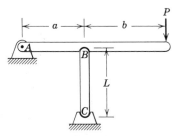

PROBLEM 12-22

12-22M *The figure shows a vertical rod, square in cross section, supporting a horizontal rod carrying the load, P, of 250 kg. Determine the required size* of the vertical rod to the nearest mm when a is 10 cm, b is 15 cm, and L is 13 cm. The rod is steel with E = 207 GPa and f_y = 345 MPa.

12-23C   The bodies H and J are absolutely rigid and immovable and their coefficient of thermal expansion is zero. The horizontal rod running between them is 1/2 in. in diameter and 30 in. long and is joined to them in a manner preventing rotation or translation. The rod is steel having an E of 29×10^6 psi, f_y of 36,000 psi, and C_t of 7.2×10^{-6} in./in./°F. Determine the increase in temperature which will cause the horizontal member to buckle.

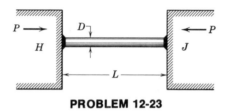

PROBLEM 12-23

12-23M   *The bodies H and J are absolutely rigid and immovable, and their coefficient of thermal expansion is zero. The horizontal rod running between them is 13 mm in diameter and 75 cm long and is joined to them in a manner preventing rotation or translation. The rod is steel having an E of 200 GPa, f_y of 248 MPa, and C_t of 1.3×10^{-5} m/m/°C.*

12-24C   The figure shows a horizontal W10X49 beam which is propped by a vertical wide-flange column. For q = 3000 lb/ft, L_b = 30 ft, L_c = 20 ft, f_y = 36,000 psi, and $E = 29 \times 10^6$ psi for the system, determine the most economical wide-flange section for the column.

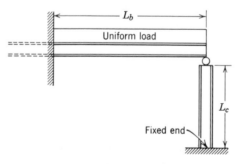

PROBLEM 12-24

12-24M *The figure shows a horizontal $250 \times 250 - 72.4\,kg/m$ wide-flange beam which is propped by a vertical wide-flange column. For $q = 4500\,kg/m$, $L_b = 9\,m$, $L_c = 6\,m$, $f_y = 248\,MPa$, and $E = 200\,GPa$ for the system, determine the most economical wide-flange section for the column.*

12-25C The figure shows a statically indeterminate beam supported by three columns. For $L = 60\,ft$, $h = 20\,ft$, the uniform load $= 275\,lb/ft$, f_y for the steel $= 36,000\,psi$, and E for the steel $29 \times 10^6\,psi$, determine the most economical wide-flange supports.

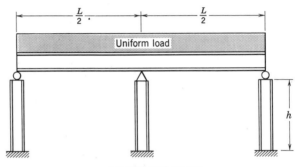

PROBLEM 12-25

12-25M *The figure shows a statically indeterminate beam supported by three columns. For $L = 18\,m$, $h = 6\,m$, the uniform load $= 400\,kg/m$, f_y for steel $= 36,000\,psi$, and E for the steel $= 200\,GPa$, determine the most economical wide-flange supports.*

APPENDIX ———————————— 1
METRIC SYSTEM—SI
THE INTERNATIONAL
SYSTEM OF UNITS

Once one becomes familiar with it, the metric system is simpler to use than the U.S. Customary System (our present system). In the past there have been a number of variations of the metric system used. For the purpose of standardizing the measuring systems of the world the international standard metric system, SI (Système International d'Unités), has been devised for use by all nations. In the following we will first describe the metric SI system, then conversion from one system to the other.

We will concern ourselves only with those metric units needed in our study of strength of materials. The *basic units* are as follows:

QUANTITY	UNIT	SYMBOL
Length	meter	m
Mass	kilogram	kg
Time	second	s
Celsius temperature	degree Celsius	°C

For convenience in handling larger or smaller quantities, prefixes representing multipliers of units are provided. Those needed for our studies are:

METRIC SYSTEM—SI THE INTERNATIONAL SYSTEM OF UNITS

MULTIPLIER	PREFIX	PREFIX SYMBOL
$1\,000\,000\,000 = 10^9$	giga	G
$1\,000\,000 = 10^6$	mega	M
$1\,000 = 10^3$	kilo	k
$0.01 = 10^{-2}$	centi	c
$0.001 = 10^{-3}$	milli	m
$0.000\,001 = 10^{-6}$	micro	μ

Note that the metric-SI system uses a *space* to set apart sets of three digits in numbers instead of the *comma* of the customary system; 1 000 000 in the metric system as compared with 1,000,000 in our customary system.

We are already familiar with the prefix *kilo* in its use in *kilo*gram. The gram of the metric system is too small to use in the majority of cases so it has been decided to use the *kilogram—one thousand* grams—as the basic unit of mass. We are also familiar with metric units as they are used in the sizes of photographic films, that is, 8 mm, 16 mm, and 32 mm where the mm stands for *millimeter*. Prefixes should always be used in preference to powers of 10 to eliminate the necessity of writing insignificant digits and decimals.

Other metric-SI units we have occasion to use are those classified as "*derived units.*" Those useful in our study are

MEASUREMENT	UNIT	SYMBOL
Area	square meter	m^2
Volume	cubic meter	m^3
Density	kilogram per cubic meter	kg/m^3
Stress	pascal (newton/square meter)	Pa (N/m^2)
Velocity	meter per second	m/s
Angular velocity	radian per second*	rad/s
Acceleration	meter per second squared	m/s^2
Angular acceleration	radian per second squared	rad/s^2
Force	newton	N (kg · m/s^2)
Energy	joule	J (N · m)
Power	watt	W (J/s)

The newton is the force required to give a mass of one kilogram an acceleration of one meter per second per second. Stated mathematically

$$1\,N = 1\,kg \cdot 1\,m/s^2$$

*The angular measure radian with its symbol rad is an S.I. unit.

In the customary system the same unit is used for force and weight. In the metric SI system the same unit is used for mass and weight.

The force of gravity acting on objects may vary by more than 5 percent from one point on the earth's surface to another. The standard acceleration g caused by the force of gravity is established at 9.80665 m/s^2. For precise measurements where g is involved the variation in the actual value of g must be taken into account. For the subject matter of this text a value of g of 9.81 is sufficiently accurate. It can therefore be stated that the force required to keep a mass of so many kilograms from falling equals that mass (in kilograms) $\times 9.81$. Thus the force of gravity acting on a mass is

$$F \text{ (in newtons)} = \text{mass (in kilograms)} \times 9.81$$

The prime source of confusion between the customary and metric systems is in the conversion of units from one system to the other. In this textbook there is no need to convert values because all examples and problems are given in either one system or the other. Realizing that, although not required in this book, you may wish to convert units from one system to the other, abbreviated conversion tables follow.

Conversion of Units

U.S. customary to metric

length

Inches (in.) $\times 2.54 \times 10^{-2}$ = *Meters (m)*
Feet (ft) $\times 0.3048$ = *Meters (m)*
Feet (ft) $\times 3.048 \times 10^{-4}$ = *Kilometers (km)*
Miles (m) $\times 1.609$ = *Kilometers (km)*

area

Square inches (in.^{2}) $\times 6.452 \times 10^{-4}$ = *Square meters (m^2)*
Square feet (ft^2) $\times 9.290 \times 10^{-2}$ = *Square meters (m^2)*
Square miles (m^2) $\times 2.589$ = *Square kilometers (km^2)*

volume

Cubic inches (in.^{3}) $\times$ 1.639 $\times$ 10^{-5} = *Cubic meters (m^3)*
Cubic feet (ft^3) $\times$ 2.832 $\times$ 10^{-2} = *Cubic meters (m^3)*
Gallons U.S. liquid (gal) $\times$ 3.785 = *Liters*

velocity

Feet per second (fps) $\times$ 0.3048 = *Meters per second (m/s)*
Feet per second (fps) $\times$ 1.097 = *Kilometers per hour (km/hr)*
Miles per hour (mph) $\times$ 1.609 = *Kilometers per hour (km/hr)*

acceleration

Feet/second/second (fps) $\times$ 0.3048 = *meters/second/second (m/s^2)*

mass

Ounces (oz) $\times$ 2.835 $\times$ 10^{-2} = *Kilograms (kg)*
Pounds (lb) $\times$ 0.4536 = *Kilograms (kg)*
Short Tons $\times$ 0.9072 = *Metric tons*

force

Pounds force (lb) $\times$ 4.448 = *Newtons (N)*
Pounds mass (lb) $\times$ 0.4536 = *Kilograms (kg)*

pressure–stress

Pounds per square inch (psi) $\times$ 6.895 $\times$ 10^3 = *Pascals (Pa)*
Pounds per square foot (lb/ft^2) $\times$ 47.88 = *Pascals (Pa)*

density

Pounds per cubic inch (lb/in.^{3})
$\times$ 2.768 $\times$ 10^4 = *Kilograms per cubic meter (kg/m^3)*
Pounds per cubic foot (lb/ft^3)
$\times$ 16.02 = *Kilograms per cubic meter (kg/m^3)*

moment

Pound-feet (lb-ft) × 1.356 = *Newton-meters (N · m)*

energy

Foot-pounds (ft-lb) × 1.356 = *Joules (J)*

momentum

Slug-feet/sec (slug-ft/sec) × 4.4482 = *Kilogram-m/sec*

power

Horsepower (Hp) × 0.7457 = *Kilowatts (kW)*
Horsepower (Hp) × 1.014 = *Metric Horsepower*

metric to U.S. customary

length

Centimeters (cm) × 0.3937 = Inches (in.)
Meters (m) × 39.37   = Inches (in.)
Meters (m) × 3.281   = Feet (ft)
Kilometers (km) × 0.6214 = Miles (m)

area

Square meters (m²) × 1550   = Square inches (in.²)
Square meters (m²) × 10.76   = Square feet (ft²)
Square Kilometers (km²) × 0.3861 = Square miles (m²)

volume

Cubic millimeters (mm³) × 6.102 × 10⁻⁵ = Cubic inches (in.³)
Cubic meters (m³) × 35.31   = Cubic feet (ft³)
Liters × 0.2642   = Gallons (gal)

velocity

Kilometers per hour (km/hr) × *0.9113* = Feet per second (fps)
Kilometers per hour (km/hr) × *0.6214* = Miles per hour (mph)

acceleration

Meters/second/second (m/s^2) × *3.281* = Feet per second per second (fps)

mass

Grams (g) × *3.527* × *10^{-2}* = Ounces (oz)
Kilograms (kg) × *35.27*　= Ounces (oz)
Kilograms (kg) × *2.205*　= Pounds (lb)
Metric tons × *1.102*　　= Short tons

force

Newtons (N) × *0.2248* = Pounds (lb)

pressure–stress

Pascals (Pa) × *1.4503* × *10^{-4}* = Pounds/square inch (psi)
Pascals (Pa) × *0.2089* × *10^{-2}* = Pounds/square foot (lb/ft^2)

density

Kilograms/cubic meter (kg/m^3) × *3.613* × *10^{-5}* = Pounds/cubic inch (lb/in.^{3})
Kilograms/cubic meter (kg/m^3) × *6.243* × *10^{-2}* = Pounds/cubic foot (lb/ft^3)

moment

Newton-meters (N·m) × *0.7375* = Pound-feet (lb-ft)

energy

Joules (J) × *0.7375* = Foot-pounds (ft-lb)

momentum

Kilogram-meters/second (kg-m/s) × *0.2248* = slug-feet/second (slug-ft/sec)

power

Kilowatts (kW) × *1.341*   = Horsepower (Hp)
Metric Horsepower × *0.9862* = Horsepower (Hp)

APPENDIX —————————— 2
PROPERTIES OF STEEL STRUCTURAL SHAPES

Inch Units

W

wide flange shapes

Designation and Nominal Size	Weight per Foot	Area	Depth	Flange		Web Thick-ness	Axis X-X			Axis Y-Y		
				Width	Thick-ness		I	S	r	I	S	r
In.	Lbs.	In.²	In.	In.	In.	In.	In.⁴	In.³	In.	In.⁴	In.³	In.
W36 ×	194	57.2	36.48	12.117	1.260	.770	12100	665	14.6	375	61.9	2.56
36 × 12	182	53.6	36.32	12.072	1.180	.725	11300	622	14.5	347	57.5	2.55
	170	50.0	36.16	12.027	1.100	.680	10500	580	14.5	320	53.2	2.53
	160	47.1	36.00	12.000	1.020	.653	9760	542	14.4	295	49.1	2.50
	150	44.2	35.84	11.972	.940	.625	9030	504	14.3	270	45.0	2.47
	135	39.8	35.55	11.945	.794	.598	7820	440	14.0	226	37.9	2.39
W33 ×	220	64.8	33.25	15.810	1.275	.775	12300	742	13.8	841	106	3.60
33 × 15¾	200	58.9	33.00	15.750	1.150	.715	11100	671	13.7	750	95.2	3.57
W33 ×	141	41.6	33.31	11.535	.960	.605	7460	448	13.4	246	42.7	2.43
33 × 11½	130	38.3	33.10	11.510	.855	.580	6710	406	13.2	218	37.9	2.38
	118	34.8	32.86	11.484	.738	.554	5900	359	13.0	187	32.5	2.32
W30 ×	116	34.2	30.00	10.500	.850	.564	4930	329	12.0	164	31.3	2.19
30 × 10½	108	31.8	29.82	10.484	.760	.548	4470	300	11.9	146	27.9	2.15
	99	29.1	29.64	10.458	.670	.522	4000	270	11.7	128	24.5	2.10

Inch Units

W

wide flange shapes

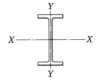

Designation and Nominal Size	Weight per Foot	Area	Depth	Flange		Web Thick-ness	Axis X-X			Axis Y-Y		
				Width	Thick-ness		I	S	r	I	S	r
In.	Lbs.	In.^{2}	In.	In.	In.	In.	In.^{4}	In.^{3}	In.	In.^{4}	In.^{3}	In.
W27 ×	94	27.7	26.91	9.990	.747	.490	3270	243	10.9	124	24.9	2.12
27 × 10	84	24.8	26.69	9.963	.636	.463	2830	212	10.7	105	21.1	2.06
W24 ×	130	38.3	24.25	14.000	.900	.565	4020	332	10.2	412	58.9	3.28
24 × 14												
W24 ×												
24 × 12	100	29.5	24.00	12.000	.775	.468	3000	250	10.1	223	37.2	2.75
W24 ×	94	27.7	24.29	9.061	.872	.516	2690	221	9.86	108	23.9	1.98
24 × 9	84	24.7	24.09	9.015	.772	.470	2370	197	9.79	94.5	21.0	1.95
	76	22.4	23.91	8.985	.682	.440	2100	176	9.69	82.6	18.4	1.92
	68	20.0	23.71	8.961	.582	.416	1820	153	9.53	70.0	15.6	1.87
W24 ×	61	18.0	23.72	7.023	.591	.419	1540	130	9.25	34.3	9.76	1.38
24 × 7	55	16.2	23.55	7.000	.503	.396	1340	114	9.10	28.9	8.25	1.34
W21 ×	55	16.2	20.80	8.215	.522	.375	1140	110	8.40	48.3	11.8	1.73
21 × 8$\frac{1}{4}$												
W21 ×	49	14.4	20.82	6.520	.532	.368	971	93.3	8.21	24.7	7.57	1.31
21 × 6$\frac{1}{2}$	44	13.0	20.66	6.500	.451	.348	843	81.6	8.07	20.7	6.38	1.27
W18 ×	70	20.6	18.00	8.750	.751	.438	1160	129	7.50	84.0	19.2	2.02
18 × 8$\frac{3}{4}$												
W18 ×	55	16.2	18.12	7.532	630	.390	891	98.4	7.42	45.0	11.9	1.67
18 × 7$\frac{1}{2}$	50	14.7	18.00	7.500	.570	.358	802	89.1	7.38	40.2	10.7	1.65
W18 ×	40	11.8	17.90	6.018	.524	.316	612	68.4	7.21	19.1	6.34	1.27
18 × 6	35	10.3	17.71	6.000	.429	.298	513	57.9	7.05	15.5	5.16	1.23
W16 ×	64	18.8	16.00	8.500	.715	.443	836	104	6.66	73.3	17.3	1.97
16 × 8$\frac{1}{2}$												
W16 ×	40	11.8	16.00	7.000	.503	.307	517	64.6	6.62	28.8	8.23	1.56
16 × 7												
W16 ×	31	9.13	15.84	5.525	.442	.275	374	47.2	6.40	12.5	4.51	1.17
16 × 5$\frac{1}{2}$	26	7.67	15.65	5.500	.345	.250	300	38.3	6.25	9.59	3.49	1.12

Inch Units

W

wide flange shapes

Designation and Nominal Size	Weight per Foot	Area	Depth	Flange		Web Thick- ness	Axis X-X			Axis Y-Y		
				Width	Thick- ness		I	S	r	I	S	r
In.	Lbs.	In.²	In.	In.	In.	In.	In.⁴	In.³	In.	In.⁴	In.³	In.
W14 × 14 × 16	730	215	22.44	17.889	4.910	3.069	14400	1280	8.18	4720	527	4.69
	665	196	21.67	17.646	4.522	2.826	12500	1150	7.99	4170	472	4.62
	605	178	20.94	17.418	4.157	2.598	10900	1040	7.81	3680	423	4.55
	264	77.6	16.50	16.025	1.938	1.205	3530	427	6.74	1330	166	4.14
	228	67.1	16.00	15.865	1.688	1.045	2940	368	6.62	1120	142	4.10
	158	46.5	15.00	15.550	1.188	.730	1900	253	6.40	745	95.8	4.00
W14 × 14 × 14½	87	25.6	14.00	14.500	.688	.420	967	138	6.15	350	48.2	3.70
W14 × 14 × 6¾	34	10.0	14.00	6.750	.453	.287	340	48.6	5.83	23.3	6.89	1.52
	30	8.83	13.86	6.733	.383	.270	290	41.9	5.74	19.5	5.80	1.49
W14 × 14 × 5	26	7.67	13.89	5.025	.418	.255	244	35.1	5.64	8.86	3.53	1.08
	22	6.49	13.72	5.000	.335	.230	198	28.9	5.53	7.00	2.80	1.04
W12 × 12 × 12	72	21.2	12.25	12.040	.671	.430	597	97.5	5.31	195	32.4	3.04
	65	19.1	12.12	12.000	.606	.390	533	88.0	5.28	175	29.1	3.02
W12 × 12 × 10	53	15.6	12.06	10.000	.576	.345	426	70.7	5.23	96.1	19.2	2.48
W12 × 12 × 8	40	11.8	11.94	8.000	.516	.294	310	51.9	5.13	44.1	11.0	1.94
W12 × 12 × 6½	27	7.95	11.96	6.497	.400	.240	204	34.2	5.07	18.3	5.63	1.52
W12 × 12 × 4	22	6.47	12.31	4.030	.424	.260	156	25.3	4.91	4.64	2.31	.847
	19	5.59	12.16	4.007	.349	.237	130	21.3	4.82	3.76	1.88	.820
	16.5	4.87	12.00	4.000	.269	.230	105	17.6	4.65	2.88	1.44	.770
W12 × 12 × 4	14	4.12	11.91	3.968	.224	.198	88.0	14.8	4.62	2.34	1.18	.754
W10 × 10 × 10	100	29.4	11.12	10.345	1.118	.685	625	112	4.61	207	39.9	2.65
	49	14.4	10.00	10.000	.558	.340	273	54.6	4.35	93.0	18.6	2.54
W10 × 10 × 8	33	9.71	9.75	7.964	.433	.292	171	35.0	4.20	36.5	9.16	1.94
W10 × 10 × 5¾	21	6.20	9.90	5.750	.340	.240	107	21.5	4.15	10.8	3.75	1.32

Inch Units

W

wide flange shapes

Designation and Nominal Size	Weight per Foot	Area	Depth	Flange		Web Thick-ness	Axis X-X			Axis Y-Y		
				Width	Thick-ness		I	S	r	I	S	r
In.	Lbs.	In.²	In.	In.	In.	In.	In.⁴	In.³	In.	In.⁴	In.³	In.
W10 × 10 × 4	19	5.61	10.25	4.020	.394	.250	96.3	18.8	4.14	4.28	2.13	.874
	15	4.41	10.00	4.000	.269	.230	68.9	13.8	3.95	2.88	1.44	.809
W10 × 10 × 4	11.5	3.39	9.87	3.950	.204	.180	52.0	10.5	3.92	2.10	1.06	.787
W8 × 8 × 8	31	9.12	8.00	8.000	.433	.288	110	27.4	3.47	37.0	9.24	2.01
W8 × 8 × 6½	24	7.06	7.93	6.500	.398	.245	82.5	20.8	3.42	18.2	5.61	1.61
W8 × 8 × 4	15	4.43	8.12	4.015	.314	.245	48.1	11.8	3.29	3.40	1.69	.876
	13	3.83	8.00	4.000	.254	.230	39.6	9.90	3.21	2.72	1.36	.842
W8 × 8 × 4	10	2.96	7.90	3.940	.204	.170	30.8	7.80	3.23	2.08	1.06	.839
W6 × 6 × 6	15.5	4.56	6.00	5.995	.269	.235	30.1	10.0	2.57	9.67	3.23	1.46
W6 × 6 × 4	12	3.54	6.00	4.000	.279	.230	21.7	7.25	2.48	2.98	1.49	.918
W6 × 6 × 4	8.5	2.51	5.83	3.940	.194	.170	14.8	5.08	2.43	1.98	1.01	.889

Inch Units

L

**angles
equal leg**

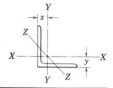

Designation and Nominal Size	Thickness	Weight per Foot	Area	Axis A-A and Axis Y-Y					Fillet Radius R
				I	*S*	*r*	*x* or *y*	r_{min}	
In.	In.	Lbs.	In.z	In.z	In.r	In.	In.	In.	In.
L8×8	$1\frac{1}{8}$	56.9	16.7	98.0	17.5	2.42	2.41	1.56	
	1	51.0	15.0	89.0	15.8	2.44	2.37	1.56	
	$\frac{7}{8}$	45.0	13.2	79.6	14.0	2.45	2.32	1.57	
	$\frac{3}{4}$	38.9	11.4	69.7	12.2	2.47	2.28	1.58	$\frac{5}{8}$
	$\frac{5}{8}$	32.7	9.61	59.4	10.3	2.49	2.23	1.58	
	$\frac{9}{16}$	29.6	8.68	54.1	9.34	2.50	2.21	1.59	
	$\frac{1}{2}$	26.4	7.75	48.6	8.36	2.50	2.19	1.59	
L6×6	1	37.4	11.0	35.5	8.57	1.80	1.86	1.17	
	$\frac{7}{8}$	33.1	9.73	31.9	7.63	1.81	1.82	1.17	
	$\frac{3}{4}$	28.7	8.44	28.2	6.66	1.83	1.78	1.17	
	$\frac{5}{8}$	24.2	7.11	24.2	5.66	1.84	1.73	1.18	
	$\frac{9}{16}$	21.9	6.43	22.1	5.14	1.85	1.71	1.18	$\frac{1}{2}$
	$\frac{1}{2}$	19.6	5.75	19.9	4.61	1.86	1.68	1.18	
	$\frac{7}{16}$	17.2	5.06	17.7	4.08	1.87	1.66	1.19	
	$\frac{3}{8}$	14.9	4.36	15.4	3.53	1.88	1.64	1.19	
	$\frac{5}{16}$	12.4	3.65	13.0	2.97	1.89	1.62	1.20	
L5×5	$\frac{7}{8}$	27.2	7.98	17.8	5.17	1.49	1.57	.973	
	$\frac{3}{4}$	23.6	6.94	15.7	4.53	1.51	1.52	.975	
	$\frac{5}{8}$	20.0	5.86	13.6	3.86	1.52	1.48	.978	
	$\frac{1}{2}$	16.2	4.75	11.3	3.16	1.54	1.43	.983	$\frac{1}{2}$
	$\frac{7}{16}$	14.3	4.18	10.0	2.79	1.55	1.41	.986	
	$\frac{3}{8}$	12.3	3.61	8.74	2.42	1.56	1.39	.990	
	$\frac{5}{16}$	10.3	3.03	7.42	2.04	1.57	1.37	.994	

PROPERTIES OF STEEL STRUCTURAL SHAPES

Designation and Nominal Size	Thick-ness	Weight per Foot	Area	Axis A-A and Axis Y-Y					Fillet Radius R
				I	S	r	x or y	r_{min}	
In.	In.	Lbs.	In.z	In.z	In.r	In.	In.	In.	In.
L4 × 4	$\frac{3}{4}$	18.5	5.44	7.67	2.81	1.19	1.27	.778	
	$\frac{5}{8}$	15.7	4.61	6.66	2.40	1.20	1.23	.779	
	$\frac{1}{2}$	12.8	3.75	5.56	1.97	1.22	1.18	.782	
	$\frac{7}{16}$	11.3	3.31	4.97	1.75	1.23	1.16	.785	$\frac{3}{8}$
	$\frac{3}{8}$	9.8	2.86	4.36	1.52	1.23	1.14	.788	
	$\frac{5}{16}$	8.2	2.40	3.71	1.29	1.24	1.12	.791	
	$\frac{1}{4}$	6.6	1.94	3.04	1.05	1.25	1.09	.795	
L3½×3½	$\frac{1}{2}$	11.1	3.25	3.64	1.49	1.06	1.06	.683	
	$\frac{3}{8}$	8.5	2.48	2.87	1.15	1.07	1.01	.687	
	$\frac{5}{16}$	7.2	2.09	2.45	.976	1.08	.990	.690	$\frac{3}{8}$
	$\frac{1}{4}$	5.8	1.69	2.01	.794	1.09	.968	.694	
L3×3	$\frac{1}{2}$	9.4	2.75	2.22	1.07	.898	.932	.584	
	$\frac{3}{8}$	7.2	2.11	1.76	.833	.913	.888	.587	
	$\frac{5}{16}$	6.1	1.78	1.51	.707	.922	.865	.589	$\frac{5}{16}$
	$\frac{1}{4}$	4.9	1.44	1.24	.577	.930	.842	.592	
	$\frac{3}{16}$	3.71	1.09	.962	.441	.939	.820	.596	

Inch Units

L

angles
unequal leg

Designation and Nominal Size	Thick-ness	Weight per Foot	Area	Axis X-X				Axis Y-Y				Axis Z-Z		Fillet Radius R
				I	S	r	y	I	S	r	x	r_{min}	Tan α	
In.	In.	Lbs.	In.²	In.⁴	In.³	In.	In.	In.⁴	In.³	In.	In.	In.		In.
L8×6	1	44.2	13.0	80.8	15.1	2.49	2.65	38.8	8.92	1.73	1.65	1.28	.543	
	7/8	39.1	11.5	72.3	13.4	2.51	2.61	34.9	7.94	1.74	1.61	1.28	.547	
	3/4	33.8	9.94	63.4	11.7	2.53	2.56	30.7	6.92	1.76	1.56	1.29	.551	
	5/8	28.5	8.36	54.1	9.87	2.54	2.52	26.3	5.88	1.77	1.52	1.29	.554	1/2
	9/16	25.7	7.56	49.3	8.95	2.55	2.50	24.0	5.34	1.78	1.50	1.30	.556	
	1/2	23.0	6.75	44.3	8.02	2.56	2.47	21.7	4.79	1.79	1.47	1.30	.558	
	7/16	20.2	5.93	39.2	7.07	2.57	2.45	19.3	4.23	1.80	1.45	1.31	.560	
L8×4	1	37.4	11.0	69.6	14.1	2.52	3.05	11.6	3.94	1.03	1.05	.846	.247	
	7/8	33.1	9.73	62.5	12.5	2.53	3.00	10.5	3.51	1.04	.999	.848	.253	
	3/4	28.7	8.44	54.9	10.9	2.55	2.95	9.36	3.07	1.05	.953	.852	.258	
	5/8	24.2	7.11	46.9	9.21	2.57	2.91	8.10	2.62	1.07	.906	.857	.262	1/2
	9/16	21.9	6.43	42.8	8.35	2.58	2.88	7.43	2.38	1.07	.882	.861	.265	
	1/2	19.6	5.75	38.5	7.49	2.59	2.86	6.74	2.15	1.08	.859	.865	.267	
	7/16	17.2	5.06	34.1	6.60	2.60	2.83	6.02	1.90	1.09	.835	.869	.269	
L7×4	7/8	30.2	8.86	42.9	9.65	2.20	2.55	10.2	3.46	1.07	1.05	.856	.318	
	3/4	26.2	7.69	37.8	8.42	2.22	2.51	9.05	3.03	1.09	1.01	.860	.324	
	5/8	22.1	6.48	32.4	7.14	2.24	2.46	7.84	2.58	1.10	.963	.865	.329	
	9/16	20.0	5.87	29.6	6.48	2.24	2.44	7.19	2.35	1.11	.940	.868	.332	1/2
	1/2	17.9	5.25	26.7	5.81	2.25	2.42	6.53	2.12	1.11	.917	.872	.335	
	7/16	15.8	4.62	23.7	5.13	2.26	2.39	5.83	1.88	1.12	.893	.876	.337	
	3/8	13.6	3.98	20.6	4.44	2.27	2.37	5.10	1.63	1.13	.870	.880	.340	
L6×4	1	30.6	9.00	30.8	8.02	1.85	2.17	10.8	3.79	1.09	1.17	.857	.414	
	7/8	27.2	7.98	27.7	7.15	1.86	2.12	9.75	3.39	1.11	1.12	.857	.421	
	3/4	23.6	6.94	24.5	6.25	1.88	2.08	8.68	2.97	1.12	1.08	.860	.428	
	5/8	20.0	5.86	21.1	5.31	1.90	2.03	7.52	2.54	1.13	1.03	.864	.435	
	9/16	18.1	5.31	19.3	4.83	1.90	2.01	6.91	2.31	1.14	1.01	.866	.438	1/2
	1/2	16.2	4.75	17.4	4.33	1.91	1.99	6.27	2.08	1.15	.987	.870	.440	
	7/16	14.3	4.18	15.5	3.83	1.92	1.96	5.60	1.85	1.16	.964	.873	.443	
	3/8	12.3	3.61	13.5	3.32	1.93	1.94	4.90	1.60	1.17	.941	.877	.446	
	5/16	10.3	3.03	11.4	2.79	1.94	1.92	4.18	1.35	1.17	.918	.882	.448	

PROPERTIES OF STEEL STRUCTURAL SHAPES

Designation and Nominal Size	Thick-ness	Weight per Foot	Area	Axis X-X				Axis Y-Y				Axis Z-Z		Fillet Radius R
				I	S	r	y	I	S	r	x	r_{min}	Tan α	
In.	In.	Lbs.	In.^{2}	In.^{4}	In.^{3}	In.	In.	In.^{4}	In.^{3}	In.	In.	In.		In.
L6×3$\frac{1}{2}$	$\frac{1}{2}$	15.3	4.50	16.6	4.24	1.92	2.08	4.25	1.59	.972	.833	.759	.344	
	$\frac{3}{8}$	11.7	3.42	12.9	3.24	1.94	2.04	3.34	1.23	.980	.787	.767	.350	$\frac{1}{2}$
	$\frac{5}{16}$	9.8	2.87	10.9	2.73	1.95	2.01	2.85	1.04	.996	.763	.772	.352	
	$\frac{1}{4}$	7.9	2.31	8.86	2.21	1.96	1.99	2.34	.847	1.01	.740	.777	.355	
L5×3$\frac{1}{2}$	$\frac{3}{4}$	19.8	5.81	13.9	4.28	1.55	1.75	5.55	2.22	.977	.996	.748	.464	
	$\frac{5}{8}$	16.8	4.92	12.0	3.65	1.56	1.70	4.83	1.90	.991	.951	.751	.472	
	$\frac{1}{2}$	13.6	4.00	9.99	2.99	1.58	1.66	4.05	1.56	1.01	.906	.755	.479	
	$\frac{7}{16}$	12.0	3.53	8.90	2.64	1.59	1.63	3.63	1.39	1.01	.883	.758	.482	$\frac{7}{16}$
	$\frac{3}{8}$	10.4	3.05	7.78	2.29	1.60	1.61	3.18	1.21	1.02	.861	.762	.486	
	$\frac{5}{16}$	8.7	2.56	6.60	1.94	1.61	1.59	2.72	1.02	1.03	.838	.766	.489	
	$\frac{1}{4}$	7.0	2.06	5.39	1.57	1.62	1.56	2.23	.830	1.04	.814	.770	.492	

Inch Units

L

angles
unequal leg

Designation and Nominal Size	Thick-ness	Weight per Foot	Area	Axis X-X				Axis Y-Y				Axis Z-Z		Fillet Radius R
				I	S	r	y	I	S	r	x	r_{min}	Tan α	
In.	In.	Lbs.	In.^{2}	In.^{4}	In.^{3}	In.	In.	In.^{4}	In.^{3}	In.	In.	In.		In.
L5 × 3	$\frac{5}{8}$	15.7	4.61	11.4	3.55	1.57	1.80	3.06	1.39	.815	.796	.644	.349	
	$\frac{1}{2}$	12.8	3.75	9.45	2.91	1.59	1.75	2.58	1.15	.829	.750	.648	.357	
	$\frac{7}{16}$	11.3	3.31	8.43	2.58	1.60	1.73	2.32	1.02	.837	.727	.651	.361	$\frac{3}{8}$
	$\frac{3}{8}$	9.8	2.86	7.37	2.24	1.61	1.70	2.04	.888	.845	.704	.654	.364	
	$\frac{5}{16}$	8.2	2.40	6.26	1.89	1.61	1.68	1.75	.753	.853	.681	.658	.368	
	$\frac{1}{4}$	6.6	1.94	5.11	1.53	1.62	1.66	1.44	.614	.861	.657	.663	.371	
L4 × 3$\frac{1}{2}$	$\frac{5}{8}$	14.7	4.30	6.37	2.35	1.22	1.29	4.52	1.84	1.03	1.04	.719	.745	
	$\frac{1}{2}$	11.9	3.50	5.32	1.94	1.23	1.25	3.79	1.52	1.04	1.00	.722	.750	
	$\frac{7}{16}$	10.6	3.09	4.76	1.72	1.24	1.23	3.40	1.35	1.05	.978	.724	.753	
	$\frac{3}{8}$	9.1	2.67	4.18	1.49	1.25	1.21	2.95	1.17	1.06	.955	.727	.755	$\frac{3}{8}$
	$\frac{5}{16}$	7.7	2.25	3.56	1.26	1.26	1.18	2.55	.994	1.07	.932	.730	.757	
	$\frac{1}{4}$	6.2	1.81	2.91	1.03	1.27	1.16	2.09	.808	1.07	.909	.734	.759	

334

Designation and Nominal Size	Thickness	Weight per Foot	Area	Axis X-X				Axis Y-Y				Axis Z-Z		Fillet Radius R
				I	S	r	y	I	S	r	x	r_{min}	Tan α	
In.	In.	Lbs.	In.^{2}	In.^{4}	In.^{3}	In.	In.	In.^{4}	In.^{3}	In.	In.	In.		In.
L4 × 3	5/8	13.6	3.98	6.03	2.30	1.23	1.37	2.87	1.35	.849	.871	.637	.534	
	1/2	11.1	3.25	5.05	1.89	1.25	1.33	2.42	1.12	.864	.827	.639	.543	
	3/8	8.5	2.48	3.96	1.46	1.26	1.28	1.92	.866	.879	.782	.644	.551	3/8
	5/16	7.2	2.09	3.38	1.23	1.27	1.26	1.65	.734	.887	.759	.647	.554	
	1/4	5.8	1.69	2.77	1.00	1.28	1.24	1.36	.599	.896	.736	.651	.558	
L3½ × 3	1/2	10.2	3.00	3.45	1.45	1.07	1.13	2.33	1.10	.881	.875	.621	.714	
	3/8	7.9	2.30	2.72	1.13	1.09	1.08	1.85	.851	.897	.830	.625	.721	3/8
	5/16	6.6	1.93	2.33	.954	1.10	1.06	1.58	.722	.905	.808	.627	.724	
	1/4	5.4	1.56	1.91	.776	1.11	1.04	1.30	.589	.914	.785	.631	.727	
L3½ × 2½	1/2	9.4	2.75	3.24	1.41	1.09	1.20	1.36	.760	.704	.705	.534	.486	
	3/8	7.2	2.11	2.56	1.09	1.10	1.16	1.09	.592	.719	.660	.537	.496	5/16
	5/16	6.1	1.78	2.19	.927	1.11	1.14	.939	.504	.727	.637	.540	.501	
	1/4	4.9	1.44	1.80	.755	1.12	1.11	.777	.412	.735	.614	.544	.506	
L3 × 2½	1/2	8.5	2.50	2.08	1.04	.913	1.00	1.30	.744	.722	.750	.520	.667	
	3/8	6.6	1.92	1.66	.810	.928	.956	1.04	.581	.736	.706	.522	.676	5/16
	5/16	5.6	1.62	1.42	.688	.937	.933	.898	.494	.744	.683	.525	.680	
	1/4	4.5	1.31	1.17	.561	.945	.911	.743	.404	.753	.661	.528	.684	
L3 × 2	1/2	7.7	2.25	1.92	1.00	.924	1.08	.672	.474	.546	.583	.428	.414	
	3/8	5.9	1.73	1.53	.781	.940	1.04	.543	.371	.559	.539	.430	.428	
	5/16	5.0	1.46	1.32	.664	.948	1.02	.470	.317	.567	.516	.432	.435	5/16
	1/4	4.1	1.19	1.09	.542	.957	.993	.392	.260	.574	.493	.435	.440	
	3/16	3.07	.902	.842	.415	.966	.970	.307	.200	.583	.470	.439	.446	

Inch Units

C

bar size channels

Designation and Nominal Size	Weight per Foot	Area	Depth d	Flanges Width b	Flanges Thickness m	Flanges Thickness n	Web Thickness w	Axis X-X I	Axis X-X S	Axis X-X r	Axis Y-Y I	Axis Y-Y S	Axis Y-Y r	Axis Y-Y x	Fillet Radius R
In.	Lbs.	In.^{2}	In.	In.	In.	In.	In.	In.^{4}	In.^{3}	In.	In.^{4}	In.^{3}	In.	In.	In.
C2$\frac{1}{2}$															
$2\frac{1}{2} \times \frac{5}{8} \times \frac{3}{16}$	2.27	.668	$2\frac{1}{2}$	$\frac{5}{8}$	$\frac{21}{64}$	$\frac{1}{8}$	$\frac{3}{16}$	.498	.399	.864	.015	.034	.151	.177	*
C2															
$2 \times 1 \times \frac{3}{16}$	2.57	.764	2	1	$\frac{1}{4}$	$\frac{7}{32}$	$\frac{3}{16}$	.428	.428	.748	.068	.102	.297	.340	$\frac{9}{64}$
C2															
$2 \times 1 \times \frac{1}{8}$	1.78	.528	2	1	$\frac{3}{16}$	$\frac{1}{8}$	$\frac{1}{8}$	.319	.319	.777	.047	.067	.297	.307	$\frac{7}{64}$
C2															
$2 \times 1 \times \frac{3}{16}$	2.32	.683	2	1	$\frac{3}{16}$	$\frac{3}{16}$	$\frac{3}{16}$	.378	.378	.744	.059	.087	.295	.317	
$2 \times 1 \times \frac{1}{8}$	1.59	.473	2	1	$\frac{1}{8}$	$\frac{1}{8}$	$\frac{1}{8}$	.279	.279	.768	.044	.062	.304	.295	$\frac{3}{32}$
C2															
$2 \times \frac{5}{8} \times \frac{1}{4}$	2.18	.641	2	$\frac{5}{8}$	$\frac{15}{64}$	$\frac{9}{64}$	$\frac{1}{4}$	.283	.283	.664	.014	.032	.147	.190	$\frac{1}{16}$
C2															
$2 \times \frac{5}{8} \times \frac{1}{4}$	2.28	.670	2	$\frac{5}{8}$	$\frac{5}{16}$	$\frac{9}{64}$	$\frac{1}{4}$	.300	.300	.669	.015	.035	.150	.198	
$2 \times \frac{9}{16} \times \frac{3}{16}$	1.86	.545	2	$\frac{9}{16}$	$\frac{5}{16}$	$\frac{9}{64}$	$\frac{3}{16}$	.258	.258	.688	.011	.028	.140	.174	$\frac{1}{32}$
$2 \times \frac{1}{2} \times \frac{1}{8}$	1.43	.420	2	$\frac{1}{2}$	$\frac{5}{16}$	$\frac{9}{64}$	$\frac{1}{8}$	.216	.216	.718	.007	.021	.133	.154	
C1$\frac{1}{2}$															
$1\frac{1}{2} \times 1\frac{1}{2} \times \frac{3}{16}$	2.72	.801	$1\frac{1}{2}$	$1\frac{1}{2}$	$\frac{13}{64}$	$\frac{3}{16}$	$\frac{3}{16}$	.274	.366	.585	.175	.188	.467	.570	$\frac{1}{8}$
C1$\frac{1}{2}$															
$1\frac{1}{2} \times \frac{3}{4} \times \frac{1}{8}$	1.17	.348	$1\frac{1}{2}$	$\frac{3}{4}$	$\frac{1}{8}$	$\frac{1}{8}$	$\frac{1}{8}$	.111	.147	.564	.016	.030	.213	.232	$\frac{3}{32}$
C1$\frac{1}{2}$															
$1\frac{1}{2} \times \frac{9}{16} \times \frac{3}{16}$	1.44	.423	$1\frac{1}{2}$	$\frac{9}{16}$	$\frac{1}{4}$	$\frac{1}{8}$	$\frac{3}{16}$	.113	.151	.518	.009	.023	.144	.180	
$1\frac{1}{2} \times \frac{1}{2} \times \frac{1}{8}$	1.12	.329	$1\frac{1}{2}$	$\frac{1}{2}$	$\frac{1}{4}$	$\frac{1}{8}$	$\frac{1}{8}$	.096	.128	.540	.006	.018	.136	.161	$\frac{1}{16}$

Metric Units

wide flange shapes

Section Index	Weight	Depth of Section	Flange Width	Thickness Web	Thickness Flange	Corner Radius	Sectional Area	Moment of Inertia I_x	Moment of Inertia I_y	Radius of Gyration r_x	Radius of Gyration r_y	Modulus of Section S_x	Modulus of Section S_y
mm	kg/m	mm	mm	mm	mm	mm	cm²	cm⁴	cm⁴	cm	cm	cm³	cm³
900 × 300	286	912	302	18	34	28	364.0	498 000	15 700	37.0	6.56	10 900	1 040
	243	900	300	16	28	28	309.8	411 000	12 600	36.4	6.39	9 140	834
	213	890	299	15	23	28	270.9	345 000	10 300	35.7	6.16	7 760	688
800 × 300	241	808	302	16	30	28	307.6	339 000	13 000	33.2	6.70	8 400	915
	210	800	300	14	26	28	267.4	292 000	11 700	33.0	6.62	7 290	782
	191	792	300	14	22	28	243.4	254 000	9 930	32.3	6.39	6 410	662
700 × 300	215	708	302	15	28	28	273.6	237 000	12 900	29.4	6.86	6 700	853
	185	700	300	13	24	28	235.5	201 000	10 800	29.3	6.78	5 760	722
	166	692	300	13	20	28	211.5	172 000	9 020	28.6	6.53	4 980	602
600 × 300	175	594	302	14	23	28	222.4	137 000	10 600	24.9	6.90	4 620	701
	151	588	300	12	20	28	192.5	118 000	9 020	24.8	6.85	4 020	601
	137	582	300	12	17	28	174.5	103 000	7 670	24.3	6.63	3 530	511
600 × 200	134	612	202	13	23	22	170.7	103 000	3 180	24.6	4.31	3 380	314
	120	606	201	12	20	22	152.5	90 400	2 720	24.3	4.22	2 980	271
	106	600	200	11	17	22	134.4	77 600	2 280	24.0	4.12	2 590	228
	94.6	596	199	10	15	22	120.5	68 700	1 980	23.9	4.05	2 310	199

Metric Units

wide flange shapes

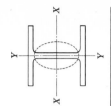

Section Index	Weight	Depth of Section	Flange Width	Thickness		Corner Radius	Sectional Area	Moment of Inertia		Radius of Gyration		Modulus of Section	
				Web	Flange			I_x	I_y	r_x	r_y	S_x	S_y
mm	kg/m	mm	mm	mm	mm	mm	cm²	cm⁴	cm⁴	cm	cm	cm³	cm³
500 × 300	128	488	300	11	18	26	163.5	71 000	8 110	20.8	7.04	2 910	541
	114	482	300	11	15	26	145.5	60 400	6 760	20.4	6.82	2 500	451
500 × 200	103	506	201	11	19	20	131.3	56 500	2 580	20.7	4.43	2 230	257
	89.7	500	200			20	114.2	47 800	2 140	20.5	4.33	1 910	214
	79.5	496	199	9	14	20	101.3	41 900	1 840	20.3	4.27	1 690	185
450 × 300	124	440	300	11	18	24	157.4	56 100	8 110	18.9	7.18	2 550	541
	106	434	299	10	15	24	135.0	46 800	6 690	18.6	7.04	2 160	448
450 × 200	76.0	450	200	9	14	18	96.76	33 500	1 870	18.6	4.40	1 490	187
	66.2	446	199	8	12	18	84.30	28 700	1 580	18.5	4.33	1 290	159
400 × 400	605	498	432	45	70	22	770.1	298 000	94 400	19.7	11.1	12 000	4 370
	415	458	417	30	50	22	528.6	187 000	60 500	18.8	10.7	8 170	2 900
	283	428	407	20	35	22	360.7	119 000	39 400	18.2	10.4	5 570	1 930
	232	414	405	18	28	22	295.4	92 800	31 000	17.7	10.2	4 480	1 530

Section		H	B	t₁	t₂		Weight	A		i_x	i_y	Z_x	Z_y
400×400	200	406	403	16	24	22	254.9	78 000	26 200	17.5	10.1	3 840	1 300
	197	400	408	21	21	22	250.7	70 900	23 800	16.8	9.75	3 540	1 170
	172	400	400	13	21	22	218.7	66 600	22 400	17.5	10.1	3 330	1 120
	168	394	405	18	18	22	214.4	59 700	20 000	16.7	9.65	3 030	985
	147	394	398	11	18	22	186.8	56 100	18 900	17.3	10.1	2 850	951
	140	388	402	15	15	22	178.5	49 000	16 300	16.6	9.54	2 520	809
400×300	107	390	300	10	16	22	136.0	38 700	7 210	16.9	7.28	1 980	481
	94.3	386	299	9	14	22	120.1	33 700	6 240	16.7	7.21	1 740	418
400×200	66.0	400	200	8	13	16	84.12	23 700	1 740	16.8	4.54	1 190	174
	56.6	396	199	7	11	16	72.16	20 000	1 450	16.7	4.48	1 010	145
350×350	159	356	352	14	22	20	202.0	47 600	16 000	15.3	8.90	2 670	909
	156	350	357	19	19	20	198.4	42 800	14 400	14.7	8.53	2 450	809
	136	350	350	12	19	20	173.9	40 300	13 600	15.2	8.84	2 300	776
	131	344	354	16	16	20	166.6	35 300	11 800	14.6	8.43	2 050	669
	115	344	348	10	16	20	146.0	33 300	11 200	15.1	8.78	1 940	646
	106	338	351	13	13	20	135.3	28 200	9 380	14.4	8.33	1 670	534
350×250	79.7	340	250	9	14	20	101.5	21 700	3 650	14.6	6.00	1 280	292
	69.2	336	249	8	12	20	88.15	18 500	3 090	14.5	5.92	1 100	248
350×175	49.6	350	175	7	11	14	63.14	13 600	984	14.7	3.95	775	112
	41.4	346	174	6	9	14	52.68	11 100	792	14.5	3.88	641	91.0
300×300	106	304	301	11	17	18	134.8	23 400	7 730	13.2	7.57	1 540	514
	106	300	305	15	15	18	134.8	21 500	7 100	12.6	7.26	1 440	466
	94.0	300	300	10	15	18	119.8	20 400	6 750	13.1	7.51	1 360	450
	87.0	298	299	9	14	18	110.8	18 800	6 240	13.0	7.51	1 270	417
	84.5	294	302	12	12	18	107.7	16 900	5 520	12.5	7.16	1 150	365

Metric Units

wide flange shapes

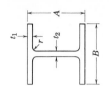

Section Index	Weight	Depth of Section (A)	Flange Width (B)	Thickness Web (t₁)	Thickness Flange (t₂)	Corner Radius (r)	Sectional Area	I_x	I_y	r_x	r_y	S_x	S_y
mm	kg/m	mm	mm	mm	mm	mm	cm²	cm⁴	cm⁴	cm	cm	cm³	cm³
300×200	65.4	298	201	9	14	18	83.36	13 300	1 900	12.6	4.77	893	189
	56.8	294	200	8	12	18	72.38	11 300	1 600	12.5	4.71	771	160
300×150	36.7	300	150	6.5	9	13	46.78	7 210	508	12.4	3.29	481	67.7
	32.0	298	149	5.5	8	13	40.80	6 320	442	12.4	3.29	424	59.3
250×250	82.2	250	255	14	14	16	104.7	11 500	3 880	10.5	6.09	919	304
	72.4	250	250	9	14	16	92.18	10 800	3 650	10.8	6.29	867	292
	66.5	248	249	8	13	16	84.70	9 930	3 350	10.8	6.29	801	269
	64.4	244	252	11	11	16	82.06	8 790	2 940	10.3	5.98	720	233
250×175	44.1	244	175	7	11	16	56.24	6 120	984	10.4	4.18	502	113
250×125	29.6	250	125	6	9	12	37.66	4 050	294	10.4	2.79	324	47.0
	25.7	248	124	5	8	12	32.68	3 540	255	10.4	2.79	285	41.1

Size													
200 × 200	65.7	208	202	10	16	13	83.69	6 530	2 200	8.83	5.13	628	218
	56.2	200	204	12	12	13	71.53	4 980	1 700	8.35	4.88	498	167
	49.9	200	200	8	12	13	63.53	4 720	1 600	8.62	5.02	472	160
200 × 150	30.6	194	150	6	9	13	39.01	2 690	507	8.30	3.61	277	67.6
200 × 100	21.3	200	100	5.5	8	11	27.16	1 840	134	8.24	2.22	184	26.8
	18.2	193	99	4.5	7	11	23.18	1 580	114	8.26	2.21	160	23.0
175 × 175	40.2	175	175	7.5	11	12	51.21	2 860	984	7.50	4.38	330	112
175 × 125	23.3	169	125	5.5	8	12	29.65	1 530	261	7.18	2.97	181	41.8
175 × 90	18.1	175	90	5	8	9	23.04	1 210	97.5	7.26	2.06	139	21.7
150 × 150	31.5	150	150	7	10	11	40.14	1 640	563	6.39	3.75	219	75.1
150 × 100	21.1	148	100	6	9	11	26.84	1 020	151	6.17	2.37	138	30.1
150 × 75	14.0	150	75	5	7	8	17.85	666	49.5	6.11	1.66	88.8	13.2
125 × 125	23.8	125	125	6.5	9	10	30.31	847	293	5.29	3.11	136	47.0
125 × 60	13.2	125	60	6	8	9	16.84	413	29.2	4.95	1.32	66.1	9.73
100 × 100	17.2	100	100	6	8	10	21.90	383	134	4.18	2.47	76.5	26.7
100 × 50	9.30	100	50	5	7	8	11.85	187	14.8	3.98	1.12	37.5	5.91

Metric Units

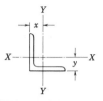

equal angles

Size mm *A × B × t*	Area of Section cm²	Weight kg/m	Center of Gravity x = y cm	Moment of Inertia $I_x = I_y$ cm⁴	Radius of Gyration $r_x = r_y$ cm	Modulus of Section $S_x = S_y$ cm³
150 × 150 × 10	29.21	22.9	4.05	627	4.63	57.3
150 × 150 × 12	34.77	27.3	4.14	740	4.61	68.2
150 × 150 × 15	42.74	33.6	4.24	888	4.56	82.6
150 × 150 × 19	53.38	41.9	4.40	1 090	4.52	103
175 × 175 × 12	40.52	31.8	4.73	1 170	5.37	91.6
175 × 175 × 15	50.21	39.4	4.85	1 440	5.35	114
200 × 200 × 15	57.75	45.3	5.47	2 180	6.14	150
200 × 200 × 20	76.00	59.7	5.67	2 820	6.09	197
200 × 200 × 25	93.75	73.6	5.87	3 420	6.04	242
200 × 200 × 29	107.6	84.5	6.01	3 866	5.99	276
250 × 250 × 25	119.4	93.7	7.10	6 950	7.63	388
250 × 250 × 35	162.6	128	7.45	9 110	7.48	519

*A = B = Length of legs
t = thickness of legs

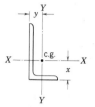

Metric Units
angles with unequal leg

Size	Sectional Area	Weight	Center of Gravity		Moment of inertia		Radius of Gyration		Modulus of Section	
*$A \times B \times t_1 \times t_2$			x	y	I_x	I_y	r_x	r_y	S_x	S_y
mm	cm²	kg/m	cm	cm	cm⁴	cm⁴	cm	cm	cm³	cm³
200 × 90 × 9 × 14	29.66	23.3	6.36	2.15	1 210	200	6.39	2.60	88.7	29.2
250 × 90 × 10 × 15	37.47	29.4	8.61	1.92	2 440	223	8.07	2.44	149	31.5
250 × 90 × 12 × 16	42.95	33.7	8.99	1.89	2 790	238	8.06	2.35	174	33.5
300 × 90 × 11 × 16	46.22	36.3	11.0	1.76	4 470	245	9.83	2.30	235	33.8
300 × 90 × 13 × 17	52.67	41.3	11.3	1.75	4 940	259	9.68	2.22	264	35.7
350 × 100 × 12 × 17	57.74	45.3	13.0	1.85	7 440	362	11.3	2.50	338	44.4
400 × 100 × 11.5 × 16	61.09	47.9	15.3	1.71	10 300	349	13.0	2.39	416	42.5
400 × 100 × 13 × 18	68.59	53.8	15.4	1.77	11 500	388	12.9	2.38	467	47.1
450 × 150 × 11.5 × 15	73.45	57.7	16.1	2.66	15 900	1 130	14.7	3.92	551	91.4
500 × 150 × 11.5 × 18	83.36	65.4	17.5	2.78	22 300	1 340	16.4	4.01	687	110
550 × 150 × 12 × 21	95.91	75.3	19.3	2.84	31 000	1 560	18.0	4.03	867	128
600 × 150 × 12.5 × 23	107.6	84.4	21.3	2.80	41 300	1 720	19.6	4.00	1 070	141

*A Length of longer leg
B Length of shorter leg
t_1 Thickness of longer leg
t_2 Thickness of shorter leg

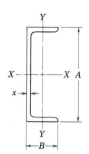

Metric Units

channels

Size mm	Sectional Area cm²	Weight kg/m	Center of Gravity cm	Moment of Inertia cm⁴		Radius of Gyration cm		Modulus of Section cm³	
$A \times B \times t_1 \times t_2$	cm²	kg/m	C_y	I_x	I_y	r_x	r_y	S_x	S_y
$300 \times 90 \times 9 \times 13$	48.57	38.1	2.23	6 440	325	11.5	2.59	429	48.0
$300 \times 90 \times 10 \times 15.5$	55.74	43.8	2.33	7 400	373	11.5	2.59	494	56.0
$300 \times 90 \times 12 \times 16$	61.90	48.6	2.25	7 870	391	11.3	2.51	525	57.9
$380 \times 100 \times 10.5 \times 16$	69.39	54.5	2.41	14 500	557	14.5	2.83	762	73.3
$380 \times 100 \times 13 \times 20$	85.71	67.3	2.50	17 600	671	14.3	2.80	924	89.5

t_1 Web thickness
t_2 Average flange thickness

344

APPENDIX ————————————————————— 3
PHYSICAL PROPERTIES OF COMMON MATERIALS

Material	Type	Strength— ksi (MPa)				Moduli ksi (MPa) $\times 10^3$		Elongation	Hardness
		Tensile	Compressive	Yield	Shear	Elastic	Shear	% in 2 in.	Brinell No.
Aluminum	2014-T6	70 (483)		60 (414)	42 (290)	10.6 (73)		13	135
Aluminum	2017-T4	62 (427)		40 (276)	38 (262)	10.5 (72)		22	105
Aluminum	2024-T4	68 (469)		47 (324)	41 (283)	10.6 (73)		19	120
Aluminum	3003-H18	29 (200)		27 (186)	16 (110)	10.0 (69)		10	55
Magnesium	AZ91C-T6	40 (276)		19 (131)	21 (145)	6.3 (43)	2.4 (16.5)	4	73
Magnesium	O1-T5	55 (379)		40 (276)	24 (165)	6.3 (43)	2.4 (16.5)	7	82
Cast Iron	40	40 (276)	120 (827)		52 (359)	17 (117)	21 (148)	0.4	220
Steel	Cold Drawn AISI C1045	103 (710)	103 (710)	90 (621)	77 (531)	30 (207)	11.4 (78.6)	14	217
Steel	Cold Drawn AISI B1112	82.5 (569)	82.5 (569)	71 (490)	62 (427)	30 (207)	11.4 (78.6)	15	170
Steel	Annealed & Cold Drawn AISI C3140	104 (717)	104 (717)	91.3 (630)	78 (538)	30 (207)	11.4 (78.6)	17	212

PHYSICAL PROPERTIES OF COMMON MATERIALS

Material	Type	Strength – ksi (MPa)				Moduli ksi (MPa) ×10³		Elongation	Hardness
		Tensile	Compressive	Yield	Shear	Elastic	Shear	% in 2 in.	Brinell No.
Steel	Hot Rolled, Annealed AISI E52100	100 (690)	100 (690)	81 (558)	75 (517)	30 (207)	11.4 (78.6)	25	192
Steel	ASTM-A36	58 (400)	58 (400)	36 (248)	43.5 (300)	29 (200)	11.9 (82)	23	
Steel	ASTM-A441	70 (483)	70 (483)	50 (345)	52.5 (362)	29 (200)	11.9 (82)	18 (in 8 inches)	
Copper	Cold Rolled 110	50 (345)		50 (276)		17 (117)		6	
#P/M Steel	F-0005-S	43 (296)		28 (193)		19 (131)		3.5	95
#P/M Steel	FC-0205-P	40 (276)		34 (234)		13 (89.6)		1.5	110
#P/M Steel	FN-0208-T	79 (545)		50 (345)		23 (159)		3.5	79
Wood	Oak	*2.05 *(14.1)	0.5 (3.45)	0.185 (1.28)	0.185 (1.28)	**1.5 (10.3)			
Plywood	Structural-1 S-1 (Dry)	*1.4 *(9.65)			0.185 (1.28)	**1.8 **(12.4)	0.90 (6.20)		

\# P/M = Powdered Metallurgy Product. * Extreme fiber in bending, ** In bending.

Note: The data in this table are for instructional use only and are not to be considered reliable for design work. For design work, information should be obtained from the manufacturers.

APPENDIX————————————————————4
COEFFICIENTS OF LINEAR EXPANSION

| | Coefficient, C_t | |
Substance	in./in./°F	m/m/°C
Aluminum	0.0000131	0.0000236
Copper	0.0000093	0.0000167
Bronze	0.0000101	0.0000182
Cast Iron	0.0000056	0.0000101
Machinery Steel	0.0000067	0.0000121
Structural Steel	0.0000072	0.0000130
Titanium	0.0000039	0.0000070
Pine Wood $\perp$ to fiber	0.0000189	0.0000341
Pine Wood $\parallel$ to fiber	0.0000030	0.0000054
Nylon	from 0.0000460	0.0000830
	to 0.0000710	0.0001280
Glass Plate	0.0000049	0.0000089

Note: The values given above may vary significantly due to variations in composition and processing. When designing, the manufacturer should be consulted concerning the value for the specific material being used.

APPENDIX———————————5
PROPERTIES OF PLANE
AREAS

Rectangle

$$I_0 = \frac{bh^3}{12} \quad I_x = \frac{bh^3}{3}$$

$$J_0 = \frac{bh}{12}(b^2 + h^2)$$

$$S_0 = \frac{bh^2}{6} \quad r = \frac{h}{\sqrt{12}}$$

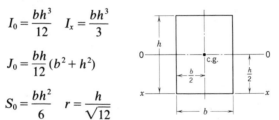

Triangle

$$I_0 = \frac{bh^3}{36} \quad I_x = \frac{bh^3}{12}$$

$$S_{\min} = \frac{bh^2}{24} \quad r = \frac{h}{\sqrt{18}}$$

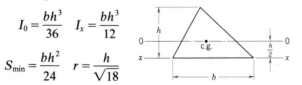

Circle

$$I_0 = \frac{\pi R^4}{4} \quad J_0 = \frac{\pi R^4}{2}$$

$$S = \frac{\pi R^3}{4} \quad r = \frac{R}{2}$$

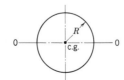

Semicircle

$$I_0 = 0.1098 \, R^4 \quad I_x = \frac{\pi \, R^4}{8}$$

$$J_0 = 0.5025 \, R^4$$

$$S_0 = 0.1907 \, R^3 \quad r = 0.2643 \, R$$

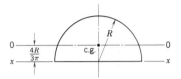

Quarter Circle

$$I_0 = 0.0549 \, R^4 \quad I_x = \frac{\pi \, R^4}{16}$$

$$J_0 = 0.1098 \, R^4$$

$$S_0 = 0.09538 \, R^3 \quad r = 0.836 \, R$$

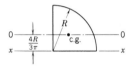

Ellipse

$$I_x = \frac{\pi a b^3}{4} \quad I_y = \frac{\pi a^3 b}{4}$$

$$J_0 = \frac{\pi a b}{4} (a^2 + b^2)$$

$$S_x = \frac{\pi a b^2}{4} \quad S_y = \frac{\pi a^2 b}{4}$$

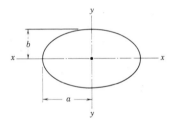

APPENDIX ———————————— 6
RECOMMENDED HOLE SIZES FOR COLD-DRIVEN SOLID RIVETS WITH CORRESPONDING SHEAR AND BEARING AREAS*

		$\frac{1}{16}$	$\frac{3}{32}$	$\frac{1}{8}$	$\frac{5}{32}$	$\frac{3}{16}$	$\frac{1}{4}$	$\frac{5}{16}$	$\frac{3}{8}$
Nominal rivet diameter, in.									
Recommended hole diameter, in.		0.067	0.096	0.1285	0.159	0.191	0.257	0.323	0.386
Corresponding drill size		51	41	30	21	11	F	P	W
Corresponding single shear area, sq. in.		0.00353	0.00724	0.01296	0.01986	0.02865	0.05187	0.08194	0.1170
		Bearing area, sq. in.							
	0.016	0.00107	------	------	------	------	------	------	------
	0.020	0.00134	0.00192	------	------	------	------	------	------
	0.025	0.00168	0.00240	0.00321	------	------	------	------	------
	0.032	0.00214	0.00307	0.00411	0.00509	------	------	------	------
	0.040	0.00268	0.00384	0.00514	0.00636	0.00764	------	------	------
	0.050	0.00335	0.00480	0.00643	0.00795	0.00955	0.01285	------	------
Sheet	0.063	0.00422	0.00605	0.00810	0.01002	0.01203	0.01619	0.0204	------
and	0.071	0.00476	0.00682	0.00912	0.01129	0.01356	0.01825	0.0229	0.0274
plate thicknesses,	0.080	------	0.00768	0.01028	0.01272	0.01528	0.02056	0.0258	0.0309
in.	0.090	------	0.00864	0.01157	0.01431	0.01719	0.02313	0.0291	0.0347
	0.100	------	0.00960	0.01285	0.01590	0.01910	0.02570	0.0323	0.0386

RECOMMENDED HOLE SIZES FOR COLD-DRIVEN SOLID RIVETS

Nominal rivet diameter, in.	$\frac{1}{16}$	$\frac{3}{32}$	$\frac{1}{8}$	$\frac{5}{32}$	$\frac{3}{16}$	$\frac{1}{4}$	$\frac{5}{16}$	$\frac{3}{8}$
Recommended hole diameter, in.	0.067	0.096	0.1285	0.159	0.191	0.257	0.323	0.386
Corresponding drill size	51	41	30	21	11	F	P	W
Corresponding single shear area, sq. in.	0.00353	0.00724	0.01296	0.01986	0.02865	0.05187	0.08194	0.1170
				Bearing area, sq. in.				
0.125	------	------	0.01606	0.01988	0.02388	0.03213	0.0404	0.0483
0.160	------	------	------	0.02544	0.03056	0.04112	0.0517	0.0618
0.190	------	------	------	------	0.03629	0.04883	0.0614	0.0733
$\frac{1}{4}$	------	------	------	------	------	0.06425	0.0808	0.0965
$\frac{5}{16}$	------	------	------	------	------	------	0.1009	0.1206
$\frac{3}{8}$	------	------	------	------	------	------	------	0.1448

Nominal rivet diameter, in.	$\frac{7}{16}$	$\frac{1}{2}$	$\frac{9}{16}$	$\frac{5}{8}$	$\frac{3}{4}$	$\frac{7}{8}$	1
Recommended hole diameter, in.	0.453	0.516	0.578	0.641	0.766	0.891	1.016
Corresponding drill size	$\frac{29}{64}$	$\frac{33}{64}$	$\frac{37}{64}$	$\frac{41}{64}$	$\frac{49}{64}$	$\frac{57}{64}$	$1\frac{1}{64}$
Corresponding single shear area, sq. in.	0.1612	0.2091	0.2624	0.3227	0.4608	0.6235	0.8107
			Bearing area, sq. in.				
0.080	0.0362	------	------	------	------	------	------
0.090	0.0408	0.0464	------	------	------	------	------
0.100	0.0453	0.0516	0.0578	------	------	------	------
0.125	0.0566	0.0645	0.0723	0.0801	0.0958	0.111	0.127
0.160	0.0725	0.0826	0.0925	0.1026	0.1226	0.143	0.163
0.190	0.0861	0.0980	0.1098	0.1218	0.1455	0.164	0.193
$\frac{1}{4}$	0.1133	0.1290	0.1445	0.1603	0.1915	0.223	0.254
$\frac{5}{16}$	0.1416	0.1613	0.1806	0.2003	0.2394	0.278	0.318
$\frac{3}{8}$	0.1689	0.1935	0.2168	0.2404	0.2873	0.334	0.381
$\frac{1}{2}$	0.2265	0.2580	0.2890	0.3205	0.3830	0.446	0.508
$\frac{5}{8}$	------	0.3225	0.3613	0.4006	0.4788	0.557	0.635
$\frac{3}{4}$	------	------	0.4335	0.4808	0.5745	0.668	0.762
1	------	------	------	0.6410	0.7660	0.891	1.016

(Left label for second table rows: Sheet and plate thicknesses, in.)

* Courtesy of Aluminum Company of America.

352

APPENDIX ——————— 7
PROPERTIES OF ALUMINUM STRUCTURAL SHAPES*

Wide-Flange Sections—H-Beams

Areas and section properties listed are based on nominal dimensions. Weights per foot are based on nominal dimensions and a density of 0.098 pounds per cubic inch, the density of alloy 6061.

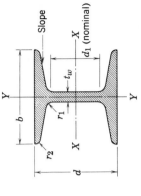

Size Nominal Depth and Width	Weight lb/ft	Actual Depth d in.	Actual Width b in.	Web Thickness t_w in.	Area in.²	Slope	Avg Flange Thickness in.	r_1 in.	r_2 in.	d_1 in.	Axis X-X I in.⁴	Axis X-X S in.³	Axis X-X r in.	Axis Y-Y I in.⁴	Axis Y-Y S in.³	Axis Y-Y r in.
4 × 4	4.76	4.00	4.000	0.313	4.05	1:11.3	0.370	0.313	0.145	2⅜	10.80	5.40	1.63	3.52	1.76	0.93
5 × 5	6.49	5.00	5.000	0.313	5.52	1:13.6	0.415	0.313	0.165	3⅜	23.94	9.58	2.08	7.73	3.09	1.18
6 × 4	4.16	6.00	4.000	0.230	3.54	0	0.279	0.250	0	4⅞	21.75	7.25	2.48	2.98	1.49	0.92
6 × 6	5.40	6.00	6.000	0.240	4.59	0	0.269	0.250	0	4⅞	30.17	10.06	2.56	9.69	3.23	1.45
6 × 6	7.85	6.00	5.933	0.250	6.68	1:15.6	0.451	0.313	0.180	4⅜	44.25	14.75	2.57	14.02	4.67	1.45
8 × 5¼	5.90	8.00	5.250	0.230	5.02	0	0.308	0.320	0	6¾	56.73	14.18	3.36	7.44	2.83	1.22
8 × 6½	8.32	8.00	6.500	0.245	7.08	0	0.398	0.400	0	6⅜	84.15	21.04	3.44	18.23	5.61	1.61
8 × 8	10.72	8.00	8.000	0.288	9.12	0	0.433	0.400	0	6⅜	109.66	27.41	3.47	36.97	9.24	2.01
8 × 8	11.24	8.00	7.938	0.313	9.55	1:18.9	0.458	0.313	0.179	6¼	113.33	28.33	3.45	33.87	8.47	1.88
8 × 8	12.99	8.00	8.125	0.500	11.05	1:18.9	0.458	0.313	0.179	6¼	121.31	30.33	3.31	36.50	9.13	1.82
10 × 5¾	7.30	9.90	5.750	0.240	6.21	0	0.340	0.312	0	8½	106.74	21.56	4.15	10.77	3.75	1.32

*Courtesy of Aluminum Company of America.

I-Beams—American Standard

Areas and section properties listed are based on nominal dimensions. Weights per foot are based on nominal and a density of 0.098 pounds per cubic inch, the density of alloy 6061.

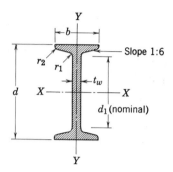

Size		Width	Web Thick-ness	Area	Avg Flange Thick-ness				Axis X-X			Axis Y-Y		
Depth d	Weight	b	t_w			r_1	r_2	d_1	I	S	r	I	S	r
in.	lb/ft	in.	in.	in.²	in.	in.	in.	in.	in.⁴	in.³	in.	in.⁴	in.³	in.
3	1.96	2.330	0.170	1.67	0.257	0.27	0.10	$1\frac{3}{4}$	2.52	1.68	1.23	0.46	0.39	0.52
3	2.59	2.509	0.349	2.21	0.257	0.27	0.10	$1\frac{3}{4}$	2.93	1.95	1.15	0.59	0.47	0.52
4	2.64	2.660	0.190	2.25	0.289	0.29	0.11	$2\frac{3}{4}$	6.06	3.03	1.64	0.76	0.57	0.58
4	3.28	2.796	0.326	2.79	0.289	0.29	0.11	$2\frac{3}{4}$	6.79	3.39	1.56	0.90	0.65	0.57
5	3.43	3.000	0.210	2.92	0.323	0.31	0.13	$3\frac{1}{2}$	12.26	4.90	2.05	1.21	0.81	0.64
5	5.10	3.284	0.494	4.34	0.323	0.31	0.13	$3\frac{1}{2}$	15.22	6.09	1.87	1.66	1.01	0.62
6	4.30	3.330	0.230	3.66	0.355	0.33	0.14	$4\frac{1}{2}$	22.08	7.36	2.46	1.82	1.09	0.71
6	5.10	3.443	0.343	4.34	0.355	0.33	0.14	$4\frac{1}{2}$	24.11	8.04	2.36	2.04	1.19	0.69
7	6.05	3.755	0.345	5.15	0.389	0.35	0.15	$5\frac{1}{4}$	39.40	11.26	2.77	2.88	1.53	0.75
8	6.35	4.000	0.270	5.40	0.421	0.37	0.16	$6\frac{1}{4}$	57.55	14.39	3.27	3.73	1.86	0.83
8	8.81	4.262	0.532	7.49	0.421	0.37	0.16	$6\frac{1}{4}$	68.73	17.18	3.03	4.66	2.19	0.79
10	8.76	4.660	0.310	7.45	0.487	0.41	0.19	8	123.39	24.68	4.07	6.78	2.91	0.95
12	10.99	5.000	0.350	9.35	0.538	0.45	0.21	$9\frac{3}{4}$	218.13	36.35	4.83	9.35	3.74	1.00

Channels—American Standard

Areas and section properties listed are based on nominal dimensions. Weights per foot are based on nominal dimensions and a density of 0.098 pounds per cubic inch, the density of alloy 6061. See ALUMINUM STANDARDS AND DATA for applicable tolerances.

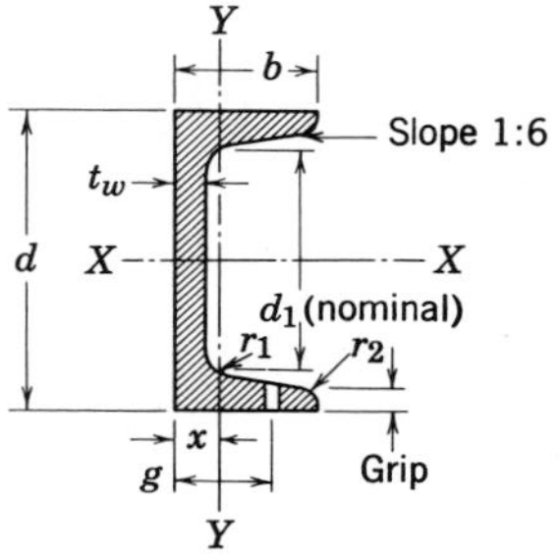

Size			Web		Avg				Axis X-X			Axis Y-Y			
Depth d in.	Weight lb/ft	Width b in.	Thickness t_w in.	Area in.^{2}	Flange Thickness in.	r_1 in.	r_2 in.	d_1 in.	I in.^{4}	S in.^{3}	r in.	I in.^{4}	S in.^{3}	r in.	x in.
3	1.42	1.410	0.170	1.20	0.270	0.27	0.10	$1\frac{3}{4}$	1.66	1.10	1.17	0.20	0.20	0.40	0.44
3	1.73	1.498	0.258	1.47	0.270	0.27	0.10	$1\frac{3}{4}$	1.85	1.24	1.12	0.25	0.23	0.41	0.44
3	2.07	1.596	0.356	1.76	0.270	0.27	0.10	$1\frac{3}{4}$	2.07	1.38	1.08	0.31	0.27	0.42	0.46
4	1.85	1.580	0.180	1.57	0.293	0.28	0.11	$2\frac{3}{4}$	3.83	1.92	1.56	0.32	0.28	0.45	0.46
4	2.16	1.647	0.247	1.84	0.293	0.28	0.11	$2\frac{3}{4}$	4.19	2.10	1.51	0.37	0.31	0.45	0.45
4	2.50	1.720	0.320	2.13	0.293	0.28	0.11	$2\frac{3}{4}$	4.58	2.29	1.47	0.43	0.34	0.45	0.46
5	2.32	1.750	0.190	1.97	0.317	0.29	0.11	$3\frac{3}{4}$	7.49	3.00	1.95	0.48	0.38	0.49	0.48
5	3.11	1.885	0.325	2.64	0.317	0.29	0.11	$3\frac{3}{4}$	8.90	3.56	1.83	0.63	0.45	0.49	0.48
5	3.97	2.032	0.472	3.38	0.317	0.29	0.11	$3\frac{3}{4}$	10.43	4.17	1.76	0.81	0.53	0.49	0.51
6	2.83	1.920	0.200	2.40	0.340	0.30	0.12	$4\frac{1}{2}$	13.12	4.37	2.34	0.69	0.49	0.54	0.51
6	3.00	1.945	0.225	2.55	0.340	0.30	0.12	$4\frac{1}{2}$	13.57	4.52	2.31	0.73	0.51	0.54	0.51
6	3.63	2.034	0.314	3.09	0.340	0.30	0.12	$4\frac{1}{2}$	15.18	5.06	2.22	0.87	0.56	0.53	0.50
6	4.48	2.157	0.437	3.82	0.340	0.30	0.12	$4\frac{1}{2}$	17.39	5.80	2.13	1.05	0.64	0.52	0.51

Size			Web		Avg				Axis X-X			Axis Y-Y			
Depth d in.	Weight lb/ft	Width b in.	Thickness t_w in.	Area in.^{2}	Flange Thickness in.	r_1 in.	r_2 in.	d_1 in.	I in.^{4}	S in.^{3}	r in.	I in.^{4}	S in.^{3}	r in.	x in.
7	3.54	2.110	0.230	3.01	0.364	0.31	0.13	$5\frac{1}{2}$	21.84	6.24	2.69	1.01	0.64	0.58	0.54
7	4.23	2.194	0.314	3.60	0.364	0.31	0.13	$5\frac{1}{2}$	24.24	6.93	2.60	1.17	0.70	0.57	0.52
7	5.10	2.299	0.419	4.33	0.364	0.31	0.13	$5\frac{1}{2}$	27.24	7.78	2.51	1.38	0.78	0.56	0.53
8	4.25	2.290	0.250	3.62	0.387	0.32	0.13	$6\frac{1}{4}$	33.85	8.46	3.06	1.40	0.81	0.62	0.56
8	4.75	2.343	0.303	4.04	0.387	0.32	0.13	$6\frac{1}{4}$	36.11	9.03	2.99	1.53	0.85	0.61	0.55
8	5.62	2.435	0.395	4.78	0.387	0.32	0.13	$6\frac{1}{4}$	40.04	10.01	2.90	1.75	0.93	0.61	0.55
8	6.48	2.527	0.487	5.51	0.387	0.32	0.13	$6\frac{1}{4}$	43.96	10.99	2.82	1.98	1.01	0.60	0.57
9	4.60	2.430	0.230	3.92	0.409	0.33	0.14	$7\frac{1}{4}$	47.68	10.60	3.49	1.75	0.96	0.67	0.60
9	6.91	2.648	0.448	5.88	0.409	0.33	0.14	$7\frac{1}{4}$	60.92	13.54	3.22	2.42	1.17	0.64	0.58
10	5.28	2.600	0.240	4.49	0.434	0.34	0.14	$8\frac{1}{4}$	67.37	13.47	3.87	2.28	1.16	0.71	0.63
10	8.64	2.886	0.526	7.35	0.434	0.34	0.14	$8\frac{1}{4}$	91.20	18.24	3.52	3.36	1.48	0.68	0.62
12	7.41	2.960	0.300	6.30	0.498	0.38	0.17	10	131.84	21.97	4.57	3.99	1.76	0.80	0.69
12	8.64	3.047	0.387	7.35	0.498	0.38	0.17	10	144.37	24.06	4.43	4.47	1.89	0.78	0.67
12	10.37	3.170	0.510	8.82	0.498	0.38	0.17	10	162.08	27.01	4.29	5.14	2.06	0.76	0.67
15	11.71	3.400	0.400	9.96	0.647	0.50	0.24	$12\frac{3}{8}$	314.76	41.97	5.62	9.63	3.11	0.90	0.79
15	17.28	3.716	0.716	14.70	0.647	0.50	0.24	$12\frac{3}{8}$	403.64	53.82	5.24	12.53	4.30	0.92	0.80

Angles with Equal Legs

Areas and section properties listed are based on nominal dimensions. Weights per foot are based on nominal dimensions and a density of 0.098 pounds per cubic inch, the density of alloy 6061.

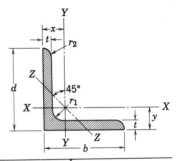

Size						Axis X-X, Y-Y			Axis Z-Z		
Depth and Width $d \times b$ in.	Weight lb/ft	Leg Thickness t in.	Area in.²	r_1 in.	r_2 in.	I in.⁴	S in.³	r in.	x, y in.	I in.⁴	r in.
$1\frac{1}{2} \times 1\frac{1}{2}$	0.43	$\frac{1}{8}$	0.36	$\frac{3}{16}$	$\frac{1}{8}$	0.074	0.068	0.45	0.41	0.031	0.29
$1\frac{1}{2} \times 1\frac{1}{2}$	0.81	$\frac{1}{4}$	0.69	$\frac{3}{16}$	$\frac{1}{8}$	0.135	0.130	0.44	0.46	0.057	0.29
$1\frac{3}{4} \times 1\frac{3}{4}$	0.58	$\frac{1}{8}$	0.42	$\frac{3}{16}$	$\frac{1}{8}$	0.121	0.094	0.53	0.47	0.050	0.34
$1\frac{3}{4} \times 1\frac{3}{4}$	0.96	$\frac{1}{4}$	0.81	$\frac{3}{16}$	$\frac{1}{8}$	0.223	0.181	0.52	0.52	0.093	0.34
2×2	0.57	$\frac{1}{8}$	0.49	$\frac{1}{4}$	$\frac{1}{8}$	0.18	0.13	0.61	0.53	0.08	0.40
2×2	1.11	$\frac{1}{4}$	0.94	$\frac{1}{4}$	$\frac{1}{8}$	0.34	0.24	0.60	0.58	0.14	0.39
2×2	1.59	$\frac{3}{8}$	1.37	$\frac{1}{4}$	$\frac{1}{8}$	0.47	0.35	0.59	0.63	0.20	0.39
$2\frac{1}{2} \times 2\frac{1}{2}$	0.72	$\frac{1}{8}$	0.62	$\frac{1}{4}$	$\frac{1}{8}$	0.37	0.20	0.77	0.65	0.15	0.50
$2\frac{1}{2} \times 2\frac{1}{2}$	1.40	$\frac{1}{4}$	1.19	$\frac{1}{4}$	$\frac{1}{8}$	0.69	0.39	0.76	0.71	0.29	0.49
$2\frac{1}{2} \times 2\frac{1}{2}$	2.05	$\frac{3}{8}$	1.71	$\frac{1}{4}$	$\frac{1}{8}$	0.98	0.56	0.75	0.76	0.41	0.48
3×3	1.68	$\frac{1}{4}$	1.43	$\frac{5}{16}$	$\frac{1}{4}$	1.18	0.54	0.91	0.82	0.49	0.58
3×3	2.47	$\frac{3}{8}$	2.10	$\frac{5}{16}$	$\frac{1}{4}$	1.70	0.80	0.90	0.87	0.70	0.58
3×3	3.23	$\frac{1}{2}$	2.77	$\frac{5}{16}$	$\frac{1}{4}$	2.16	1.04	0.89	0.92	0.91	0.58
$3\frac{1}{2} \times 3\frac{1}{2}$	1.99	$\frac{1}{4}$	1.69	$\frac{3}{8}$	$\frac{1}{4}$	1.93	0.76	1.07	0.94	0.80	0.69
$3\frac{1}{2} \times 3\frac{1}{2}$	2.46	$\frac{5}{16}$	2.09	$\frac{3}{8}$	$\frac{1}{4}$	2.37	0.94	1.06	0.97	0.98	0.68
$3\frac{1}{2} \times 3\frac{1}{2}$	2.93	$\frac{3}{8}$	2.49	$\frac{3}{8}$	$\frac{1}{4}$	2.79	1.11	1.06	1.00	1.15	0.68
$3\frac{1}{2} \times 3\frac{1}{2}$	3.83	$\frac{1}{2}$	3.25	$\frac{3}{8}$	$\frac{1}{4}$	3.56	1.45	1.05	1.05	1.49	0.68
4×4	2.28	$\frac{1}{4}$	1.94	$\frac{3}{8}$	$\frac{1}{4}$	2.94	1.00	1.23	1.07	1.21	0.79
4×4	2.83	$\frac{5}{16}$	2.41	$\frac{3}{8}$	$\frac{1}{4}$	3.61	1.24	1.23	1.10	1.48	0.78
4×4	3.38	$\frac{3}{8}$	2.86	$\frac{3}{8}$	$\frac{1}{4}$	4.26	1.48	1.22	1.12	1.75	0.78

PROPERTIES OF ALUMINUM STRUCTURAL SHAPES

| Size | | | | | | Axis X-X, Y-Y | | | | Axis Z-Z | |
Depth and Width $d \times b$ in.	Weight lb/ft	Leg Thickness t in.	Area in.^{2}	r_1 in.	r_2 in.	I in.^{4}	S in.^{3}	r in.	x, y in.	I in.^{4}	r in.
4×4	3.90	$\frac{7}{16}$	3.31	$\frac{3}{8}$	$\frac{1}{4}$	4.87	1.71	1.21	1.15	2.01	0.78
4×4	4.41	$\frac{1}{2}$	3.75	$\frac{3}{8}$	$\frac{1}{4}$	5.46	1.93	1.21	1.17	2.26	0.78
4×4	4.92	$\frac{9}{16}$	4.18	$\frac{3}{8}$	$\frac{1}{4}$	6.02	2.15	1.20	1.20	2.51	0.77
4×4	5.42	$\frac{5}{8}$	4.61	$\frac{3}{8}$	$\frac{1}{4}$	6.56	2.36	1.19	1.22	2.76	0.77
4×4	5.92	$\frac{11}{16}$	5.03	$\frac{3}{8}$	$\frac{1}{4}$	7.08	2.57	1.19	1.24	3.00	0.77
4×4	6.40	$\frac{3}{4}$	5.44	$\frac{3}{8}$	$\frac{1}{4}$	7.57	2.77	1.18	1.26	3.25	0.77
5×5	4.28	$\frac{3}{8}$	3.60	$\frac{1}{2}$	$\frac{3}{8}$	8.37	2.30	1.52	1.36	3.44	0.98
5×5	4.91	$\frac{7}{16}$	4.12	$\frac{1}{2}$	$\frac{3}{8}$	9.65	2.67	1.52	1.38	3.96	0.97
5×5	5.58	$\frac{1}{2}$	4.74	$\frac{1}{2}$	$\frac{3}{8}$	10.89	3.03	1.52	1.41	4.47	0.97
5×5	6.88	$\frac{5}{8}$	5.85	$\frac{1}{2}$	$\frac{3}{8}$	13.22	3.73	1.50	1.46	5.47	0.97
6×6	5.12	$\frac{3}{8}$	4.35	$\frac{1}{2}$	$\frac{3}{8}$	14.85	3.38	1.85	1.60	6.07	1.18
6×6	5.94	$\frac{7}{16}$	5.05	$\frac{1}{2}$	$\frac{3}{8}$	17.15	3.93	1.84	1.63	7.01	1.18
6×6	6.75	$\frac{1}{2}$	5.74	$\frac{1}{2}$	$\frac{3}{8}$	19.38	4.46	1.84	1.66	7.92	1.17
6×6	8.35	$\frac{5}{8}$	7.10	$\frac{1}{2}$	$\frac{3}{8}$	23.64	5.51	1.82	1.71	9.70	1.17
8×8	9.14	$\frac{1}{2}$	7.77	$\frac{5}{8}$	$\frac{3}{8}$	47.74	8.16	2.48	2.15	19.51	1.58
8×8	13.48	$\frac{3}{4}$	11.46	$\frac{5}{8}$	$\frac{3}{8}$	68.86	11.99	2.45	2.26	28.20	1.57
8×8	17.67	1	15.02	$\frac{5}{8}$	$\frac{3}{8}$	88.11	15.60	2.42	2.35	36.46	1.56

Angles with Unequal Legs

Areas and section properties listed are based on nominal dimensions. Weights per foot are based on nominal dimensions and a density of 0.098 pounds per cubic inch, the density of alloy 6061.

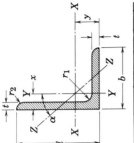

Size — Depth and Width d×b in.	Weight lb/ft	Leg Thickness t in.	Area in.²	r₁ in.	r₂ in.	Axis X-X I in.⁴	Axis X-X S in.³	Axis X-X r in.	Axis X-X y in.	Axis Y-Y I in.⁴	Axis Y-Y S in.³	Axis Y-Y r in.	Axis Y-Y x in.	Axis Z-Z α	Axis Z-Z I in.⁴	Axis Z-Z r in.
*1¾×1¼	0.42	1/8	0.36	3/16	1/8	0.108	0.090	0.55	0.54	0.046	0.048	0.36	0.30	26°22'	0.026	0.27
*1¾×1¼	0.81	1/4	0.69	3/16	1/8	0.199	0.172	0.54	0.60	0.083	0.092	0.35	0.35	25°47'	0.048	0.26
*2×1½	0.50	1/8	0.42	3/16	1/8	0.17	0.12	0.63	0.60	0.08	0.07	0.44	0.36	28°44'	0.04	0.32
*2×1½	0.96	1/4	0.81	3/16	1/8	0.31	0.23	0.62	0.66	0.15	0.14	0.43	0.41	28°20'	0.08	0.32
*2×1½	1.38	3/8	1.17	3/16	1/8	0.43	0.33	0.60	0.70	0.20	0.19	0.41	0.45	27°37'	0.12	0.32
*2½×1½	1.11	1/4	0.94	1/4	1/8	0.59	0.36	0.79	0.86	0.16	0.14	0.41	0.37	19°25'	0.10	0.32
*2½×2	0.65	1/8	0.55	1/4	1/8	0.34	0.19	0.79	0.72	0.20	0.13	0.60	0.48	31°57'	0.10	0.43
*2½×2	1.26	1/4	1.07	1/4	1/8	0.65	0.38	0.78	0.78	0.37	0.25	0.58	0.53	31°51'	0.19	0.42
*2½×2	1.83	3/8	1.55	1/4	1/8	0.91	0.54	0.76	0.83	0.51	0.36	0.57	0.58	31°28'	0.27	0.42
*3×2	1.40	1/4	1.19	5/16	3/16	1.06	0.52	0.94	0.97	0.38	0.25	0.56	0.48	23°22'	0.22	0.43
*3×2	2.05	3/8	1.74	5/16	3/16	1.51	0.76	0.93	1.03	0.53	0.36	0.55	0.53	23°0'	0.31	0.42
*3×2½	1.54	1/4	1.31	5/16	1/4	1.12	0.53	0.92	0.89	0.70	0.38	0.73	0.64	34°03'	0.35	0.52
*3×2½	2.25	3/8	1.92	5/16	1/4	1.60	0.78	0.91	0.94	1.00	0.55	0.72	0.69	33°54'	0.51	0.51
*3½×2½	1.68	1/4	1.43	5/16	1/4	1.73	0.72	1.10	1.09	0.73	0.38	0.71	0.60	26°23'	0.41	0.53
*3½×2½	2.08	5/16	1.77	5/16	1/4	2.12	0.89	1.09	1.12	0.89	0.48	0.71	0.62	26°18'	0.50	0.53
*3½×2½	2.47	3/8	2.10	5/16	1/4	2.49	1.06	1.09	1.14	1.05	0.57	0.71	0.65	26°10'	0.59	0.53
*3½×2½	3.23	1/2	2.74	5/16	1/4	3.17	1.37	1.08	1.19	1.32	0.73	0.69	0.70	25°48'	0.76	0.53
*3½×3	1.84	1/4	1.57	3/8	1/4	1.84	0.74	1.08	1.01	1.28	0.57	0.90	0.76	36°13'	0.63	0.63
*3½×3	2.28	5/16	1.94	3/8	1/4	2.26	0.92	1.08	1.04	1.52	0.69	0.89	0.79	35°40'	0.74	0.62
*3½×3	2.70	3/8	2.30	3/8	1/4	2.65	1.09	1.07	1.06	1.79	0.82	0.88	0.82	35°37'	0.87	0.62
*3½×3	3.53	1/2	3.00	3/8	1/4	3.38	1.42	1.06	1.11	2.27	1.06	0.87	0.86	35°26'	1.13	0.61

Size						Axis X-X				Axis Y-Y				Axis Z-Z		
Depth and Width d×b in.	Weight lb/ft	Leg Thickness t in.	Area in.²	r_1 in.	r_2 in.	I in.⁴	S in.³	r in.	y in.	I in.⁴	S in.³	r in.	x in.	α	I in.⁴	r in.
*4×3	1.99	1/4	1.69	3/8	1/4	2.68	0.96	1.26	1.21	1.29	0.56	0.87	0.72	28°42'	0.70	0.64
*4×3	2.46	5/16	2.09	3/8	1/4	3.29	1.19	1.25	1.24	1.58	0.70	0.87	0.74	28°40'	0.85	0.64
*4×3	2.93	3/8	2.49	3/8	1/4	3.88	1.42	1.25	1.26	1.86	0.83	0.86	0.77	28°35'	1.01	0.64
*4×3	3.38	7/16	2.87	3/8	1/4	4.43	1.63	1.24	1.29	2.12	0.96	0.86	0.79	28°28'	1.15	0.63
*4×3	3.83	1/2	3.25	3/8	1/4	4.96	1.85	1.24	1.31	2.36	1.08	0.85	0.82	28°20'	1.30	0.63
*4×3	4.26	9/16	3.62	3/8	1/4	5.47	2.05	1.23	1.34	2.60	1.20	0.85	0.84	28°11'	1.44	0.63
*4×3½	3.13	3/8	2.66	3/8	5/16	4.02	1.43	1.23	1.18	2.85	1.11	1.04	0.94	36°53'	1.36	0.72
*4×3½	4.10	1/2	3.49	3/8	5/16	5.17	1.87	1.22	1.23	3.67	1.46	1.03	0.99	36°46'	1.77	0.71
*5×3	3.35	3/8	2.85	3/8	5/16	7.15	2.15	1.59	1.68	1.93	0.84	0.82	0.69	19°40'	1.17	0.64
*5×3	4.40	1/2	3.74	3/4	5/16	9.24	2.83	1.57	1.73	2.48	1.10	0.81	0.74	19°26'	1.52	0.64
*5×3½	3.01	5/16	2.57	7/16	5/16	6.39	1.85	1.58	1.55	2.58	0.96	1.00	0.81	25°33'	1.45	0.75
*5×3½	3.58	3/8	3.05	7/16	5/16	7.56	2.21	1.58	1.58	3.04	1.15	1.00	0.84	25°32'	1.71	0.75
*5×3½	4.15	7/16	3.53	7/16	5/16	8.69	2.56	1.57	1.61	3.49	1.33	0.99	0.87	25°27'	1.97	0.75
*5×3½	4.70	1/2	4.00	7/16	5/16	9.77	2.90	1.56	1.63	3.91	1.50	0.99	0.89	25°22'	2.22	0.74
*5×3½	5.79	5/8	4.92	7/16	5/16	11.82	3.56	1.55	1.68	4.70	1.84	0.98	0.94	25°07'	2.70	0.74
*6×3½	3.39	5/16	2.88	1/2	5/16	10.64	2.64	1.92	1.97	2.70	0.98	0.97	0.74	18°52'	1.65	0.76
*6×3½	4.04	3/8	3.43	1/2	5/16	12.60	3.15	1.92	2.00	3.19	1.17	0.96	0.77	18°51'	1.95	0.75
*6×3½	5.31	1/2	4.51	1/2	5/16	16.34	4.14	1.90	2.06	4.11	1.53	0.95	0.82	18°42'	2.52	0.75
*6×4	4.24	3/8	3.60	1/2	3/8	13.02	3.17	1.90	1.90	4.63	1.50	1.13	0.91	23°33'	2.67	0.86
*6×4	4.91	7/16	4.18	1/2	3/8	15.02	3.69	1.90	1.93	5.34	1.74	1.13	0.94	23°31'	3.07	0.86
*6×4	5.58	1/2	4.74	1/2	3/8	16.95	4.19	1.89	1.96	6.01	1.98	1.13	0.97	23°27'	3.47	0.86
*6×4	6.24	9/16	5.30	1/2	3/8	18.82	4.69	1.88	1.98	6.65	2.21	1.12	0.99	23°22'	3.86	0.85
*6×4	6.88	5/8	5.85	1/2	3/8	20.63	5.17	1.88	2.01	7.27	2.44	1.11	1.02	23°16'	4.24	0.85
*6×4	8.15	3/4	6.93	1/2	3/8	24.08	6.11	1.86	2.06	8.43	2.87	1.10	1.07	23°02'	4.98	0.85
*8×6	9.84	5/8	8.37	1/2	5/16	53.57	9.74	2.53	2.50	25.94	5.77	1.76	1.50	28°52'	13.89	1.29
*8×6	10.76	11/16	9.15	1/2	3/8	57.99	10.58	2.52	2.52	27.98	6.25	1.75	1.53	28°47'	15.02	1.28
*8×6	11.68	3/4	9.93	1/2	3/8	62.60	11.47	2.51	2.54	30.15	6.77	1.74	1.55	28°43'	16.24	1.28

Answers to Even-Numbered Problems

chapter 2

2-2C	156 psi
2-2M	*0.981 MPa*
2-4C	12,800 psi
2-4M	*79.7 MPa*
2-6C	32,600 psi
2-6M	*554 MPa*
2-8C	W10X15
2-8M	*350 × 175 − 49.6 kg/m*
2-10C	288 ft
2-10M	*87.9 m*
2-12C	2.27 in.
2-12M	*37.8 mm*
2-14C	45 in.
2-14M	*2.50 m*
2-16C	210 psi
2-16M	*420 Pa*
2-18C	30,000,000 psi
2-18M	*212 GPa*
2-20C	0.0640 in.
2-20M	$1.64 \times 10^{-3}\,m$
2-22C	(a) W12X40, (b) 0.166
2-22M	*(a) 200 × 200 − 56.2 kg/m, (b) 4.12 mm*
2-24C	$5.41 \times 10^{-3}\,in.$
2-24M	$1.89 \times 10^{-4}\,m$
2-26C	1.18 in., 18.4°
2-26M	*30.3 mm, 18.3°*
2-28C	0.299
2-28M	*0.299*
2-30C	1.999406 in. on a side.
2-30M	*0.049 985 m on a side.*

ANSWERS TO EVEN-NUMBERED PROBLEMS

2-32C	0.0000075
2-32M	*0.000 016 2*
2-34C	10.8°F
2-34M	*5.99°C*
2-36C	61,900 lb
2-36M	*817 kN*
2-38C	120.08 in.
2-38M	*3.001 56 m*
2-40C	25,600 psi
2-40M	*178 MPa*
2-42C	1.03, controlling member is *CD*.
2-42M	*1.03, controlling member is AD.*

chapter 3

3-2C	0.0415 in.
3-2M	*0.99 mm*
3-4C	11,500 lb
3-4M	*49 300 N*
3-6C	3520 lb
3-6M	*14 100 N*
3-8C	$f_v = 1330$ psi, $\varepsilon = 1.33 \times 10^{-4}$ in./in., $\Delta = 6.67 \times 10^{-5}$ in.
3-8M	*$f_v = 17.8$ MPa, $\varepsilon = 2.54 \times 10^{-4}$ m/m, $\Delta = 3.81 \times 10^{-6}$ m*
3-10C	1980 psi
3-10M	*14.3 MPa*
3-12C	$f_v \, 44° = 0.4997 \, f_a$; $f_v \, 45° = 0.5000 \, f_a$ (maximum); $f_v \, 46° = 0.4997 \, f_a$.
3-12M	*Same as 3-12C.*
3-14C	669 lb
3-14M	*2 870 N*
3-16C	5.8 in.
3-16M	*14.5 cm*
3-18C	$\Delta_a = 0.00116$ in., $\Delta_b = 0.00116$ in., $f_a = 933$ psi, $f_b = 561$ psi.

3-18M $\Delta_a = 3.57 \times 10^{-5}\,m$, $\Delta_b = 3.57 \times 10^{-5}\,m$, $f_a = 7\,114\,MPa$, $f_b = 4.46\,MPa$.

3-20C   319°F, 239.782 in.

3-20M   117°C, 5.994 53 m

3-22C   Rod 1, $1\frac{7}{32}$ in. dia., rod 2, $1\frac{13}{32}$ in. dia.

3-22M   Rod 1, 45 mm dia., rod 2, 50 mm dia.

3-24C   3390 in.-lb

3-24M   214 J

3-26C   Steel, 22.3 in.-lb/in.^{3}; aluminum, 171 in.-lb/in.^{3}

3-26M   Steel, 156 kJ/m^3; aluminum, 1.17 MJ/m^3.

3-28C   89,100 psi

3-28M   409 MPa

3-30C   0.479 lb

3-30M   0.240 kg

3-32C   520,000 psi

3-32M   3.50 GPa

chapter 4

4-2C   375 psi

4-2M   2.67 MPa

4-4C   0.372 in.

4-4M   8.9 mm

4-6C   5400 psi

4-6M   31.8 MPa

4-8C   1.30 in.

4-8M   34.5 mm

4-10C   147 in.

4-10M   4.08 m

4-12C   481 ft or less.

4-12M   136 m or less.

chapter 5

5-2C   Failure by crushing behind rivets at 787 lb.

5-2M   Failure by crushing behind rivets at 3 050 N.

ANSWERS TO EVEN-NUMBERED PROBLEMS

5-4C f_v = 1440 psi, f_c = 2920 psi, f_t(row 1 or 2) = 1210 psi.

5-4M   *f_v = 12.1 MPa, f_c = 24 MPa, f_t(row 1 or 2) = 8.87 MPa.*

5-6C   16,500 lb

5-6M   *56 kN*

5-8C f_{vb} = 6240 psi, f_{cm} = 17,100 psi, f_{cc} = 13,700 psi,
f_t(row 1-main) = 60,000 psi, f_t(row 2-main) = 25,000 psi
f_t(row 3-main) = 22,500 psi, f_t(row 1-cover) = 18,000 psi,
f_t(row 2-cover) = 20,000 psi, f_t(row 3-cover) = 48,000 psi.

5-8M   *f_{vb} = 49.8 MPa, f_{cm} = 104 MPa, f_{cc} = 78 MPa,*
f_t(row 1-main) = 301 MPa, f_t(row 2-main) = 132 MPa
f_t(row 3-main) = 113 MPa, f_t(row 1-cover) = 84.6 MPa
f_t(row 2-cover) = 99.4 MPa, f_t(row 3-cover) = 226 MPa.

5-10C   95.4 kip

5-10M   *464 kN*

5-12C   0.407, use 7/16 bolts.

5-12M   *0.008 26 m, use 10 mm bolts.*

5-14C   1.19 in., use $1\frac{1}{8}$ in. rivets.

5-14M   *30.5 mm, use 36 mm rivets.*

5-16C   36,000 lb

5-16M   *161 kN*

5-18C   3970 psi

5-18M   *29.5 MPa*

5-20C   7.86 in., use 4 in. on a side.

5-20M   *157 mm, use 8 cm on a side.*

5-22C   76.4 kips

5-22M   *254 kN*

5-24C   6 in. on corner, $2\frac{5}{8}$ in. on edge.

5-24M   *15 cm on corner, 6 cm on edge.*

5-26C $4\frac{5}{8}$ in. on corner, $1\frac{1}{8}$ in. on edge.

5-26M   *12 cm on corner, 3 cm on edge.*

5-28C $4\frac{1}{4}$ in. on right, $2\frac{1}{8}$ in. on left.

5-28M   *14 cm on right, 7 cm on left.*

5-30C	$3\frac{1}{2}$ in. on left, $6\frac{1}{4}$ in. on right.
5-30M	*7.5 cm on left, 13 cm on right.*

chapter 6

6-2C	2.39 in., use $2\frac{7}{16}$ in.
6-2M	*61.2 mm, use 62 mm*
6-4C	9630 lb-in.
6-4M	*929 N · m*
6-6C	0.15 in.
6-6M	*4.10 mm*
6-8C	6420 psi
6-8M	*30 MPa*
6-10C	0.894 in.
6-10M	*44 mm*
6-12C	299 HP
6-12M	*31.3 kW*
6-14C	Use 5 bolts.
6-14M	*Use 5 bolts.*
6-16C	428,000 lb-in.
6-16M	*42 400 N · m*
6-18C	0.0386 rad = 2.21°
6-18M	*0.389 rad = 2.23°*
6-20C	4.20 in.
6-20M	*178 mm*
6-22C	0.227 rad = 13°
6-22M	*0.239 rad = 13.7°*
6-24C	744 in.-lb.
6-24M	*5.25 J*

chapter 7

Problems for this chapter require drawing of shear and moment diagrams. Therefore, there are no answers.

ANSWERS TO EVEN-NUMBERED PROBLEMS

chapter 8

8-2C	1190 lb
8-2M	*584 kg*
8-4C	669 psi
8-4M	*4.55 MPa*
8-6C	40,100 lb
8-6M	*18 000 kg*
8-8C	h = 10 in., b = 5 in.
8-8M	*h = 260 mm, b = 130 mm*
8-10C	$P_{1,\,max} = 62.6$ lb, $P_{2,\,max} = 20.9$ lb
8-10M	$P_{1,\,max} = 26.7$ kg, $P_{2,\,max} = 10$ kg
8-12C	2930 lb/ft
8-12M	*1360 kg/m*
8-14C	21.1 psi
8-14M	*73.5 kPa*
8-16C	2350 lb
8-16M	*1250 kg*
8.18C	45 psi, satisfactory.
8-18M	*327 kPa, satisfactory.*
8-20C	Unsafe.
8-20M	*Safe.*
8-22C	143 psi
8-22M	*5.49 MPa*
8-24C	104 psi
8-24M	*736 kPa*
8-26C	9.91 psi
8-26M	*420 kPa*
8-28C	20.4 lb/ft
8-28M	*14.1 kg/m*
8-30C	278 lb/ft
8-30M	*438 kg/m*

chapter 9

9-2C	3.29°
9-2M	*3.30°*
9-4C	0.465°
9-4M	*0.445°*
9-6C	4.64 in.
9-6M	*0.174 m*
9-8C	0.795 in.
9-8M	*0.193 m*
9-10C	0.800 in.
9-10M	*30.4 mm*
9.12C	360 lb
9-12M	*1270 kg*
9-14C	0.606 in.
9-14M	*16.2 mm*
9-16C	3.61 in.
9-16M	*0.104 m*
9-18C	0.149 in. at 132 in. from right end.
9-18M	*4.48 mm at 3.26 m from right end.*
9-20C	0.0698 in. at 155 in. from left end.
9-20M	*1.8 mm at 3.94 m from left end.*
9-22C	0.392 in.
9-22M	*0.01 m*
9-24C	0.122 in.
9-24M	*3.07 mm*
9-26C	77.8 lb/ft
9-26M	*124 kg/m*
9-28C	1.73 in.
9-28M	*41.1 mm*
9-30C	5.34 in.
9-30M	*138 mm*
9-32C	6.23 in.

ANSWERS TO EVEN-NUMBERED PROBLEMS

9-32M	*205 mm*
9-34C	0.913 in.
9-34M	*53.3 mm*
9-36C	1.43 in.
9-36M	*32.7 mm*
9-38C	0.926 in.
9-38M	*63.8 mm*
9-40C	2.00 in.
9-40M	*0.051 9 m*

chapter 10

10-2C	$R_A = 5620$ lb, $R_B = 3380$ lb, $M_A = 20,200$ lb-ft
10-2M	$R_A = 27.5\,kN$, $R_B = 16.7\,kN$, $M_A = 33\,kN \cdot m$
10-4C	$R_A = 963$ lb, $R_B = 1040$ lb, $M_A = 5330$ lb-ft
10-4M	$R_A = 4.72\,kN$, $R_B = 5.09\,kN$, $M_A = 8.72\,kN \cdot m$
10-6C	$R_A = -3750$ lb, $R_B = 8750$ lb, $M_A = -7500$ lb-ft
10-6M	$R_A = -55.2\,kN$, $R_B = 129\,kN$, $M_A = -36.8\,kN \cdot m$
10-8C	$R_A = -375$ lb, $R_B = 6370$ lb, $M_A = -3750$ lb-ft
10-8M	$R_A = -1.64\,kN$, $R_B = 28.1\,kN$, $M_A = -4.92\,kN \cdot m$
10-10C	$R_A = -350$ lb, $R_B = 3950$ lb, $M_A = -1300$ lb
10-10M	$R_A = -0.124\,kN$, $R_b = 21.7\,kN$, $M_A = -2.32\,N \cdot m$
10-12C	$R_A = 9000$ lb, $R_B = 13,000$ lb, $M_A = 39,000$ lb-ft
10-12M	$R_A = 39.9\,kN$, $R_B = 57.8\,kN$, $M_A = 57.5\,kN \cdot m$
10-14C	$R_A = R_B = 25$ lb, $M_A = M_B = 4.17$ lb-ft
10-14M	$R_A = R_B = 110\,N$, $M_A = M_B = 5.52\,N \cdot m$
10-16C	$R_A = 6150$ lb, $R_B = 7850$ lb, $M_A = 420,000$ lb-ft, $M_B = 375,000$ lb-ft
10-16M	$R_A = 28.3\,kN$, $R_B = 40.4\,kN$, $M_A = 32.7\,kN \cdot m$, $M_B = 44.4\,kN \cdot m$
10-18C	$R_A = 262$ lb, $R_B = 676$ lb, $M_A = 597$ lb-ft
10-18M	$R_A = 1200\,N$, $R_B = 3000\,N$, $M_A = 836\,N \cdot m$
10-20C	$R_A = 965$ lb, $R_B = 34.9$ lb, $R_C = 34.9$ lb, $M_A = 2330$ lb-ft, $M_C = 349$ lb-ft
10-20M	$R_A = 4.26\,kN$, $R_B = 0.152\,kN$, $R_C = 0.152\,kN$, $M_A = 3.14\,kN \cdot m$, $M_C = 0.456\,kN \cdot m$

10-22C	$R_A = R_C = 2500$ lb, $R_B = 11,000$ lb, $M_B = -22,500$ lb-ft
10-22M	$R_A = R_C = 10.7$ kN, $R_B = 47.2$ kN, $M_B = -29$ kN · m
10-24C	$R_A = 790$ lb, $R_B = 2270$ lb, $R_C = 538$ lb, $M_B = -2100$ lb-ft
10-24M	$R_A = 3.46$ kN, $R_B = 10.2$ kN, $R_C = 2.53$ kN, $M_B = -2.86$ kN · m
10-26C	$R_A = 1020$ lb, $R_B = 10,500$ lb, $R_C = 8520$ lb, $M_B = -14,800$ lb-ft, $M_C = 50,000$ lb-ft
10-26M	$R_A = 4.07$ kN, $R_B = 41.9$ kN, $R_C = 34.1$ kN, $M_B = -17.8$ kN · m, $M_C = 60$ kN · m
10-28C	$R_A = 4800$ lb, $R_B = 8080$ lb, $R_C = 5200$ lb, $R_D = 2420$ lb
10-28M	$R_A = 21.2$ kN, $R_B = 35.2$ kN, $R_C = 11.2$ kN, $R_D = 11.2$ kN

chapter 11

	$f_{n,\,max}$	$f_{v,\,max}$
11-2C	20,000 psi	10,000 psi
11-2M	140 MPa	70 MPa
11-4C	22,100 psi	7070 psi
11-4M	154 MPa	49.5 MPa
11-6C	20,000 psi	10,000 psi
11-6M	140 MPa	70 MPa
11-8C	10,000 psi	10,000 psi
11-8M	70 MPa	70 MPa
11-10C	20,000 psi	5000 psi
11-10M	140 MPa	35 MPa
11-12C	-1110 psi	853 psi
11-12M	-7.58 MPa	5.58 MPa
11-14C	$f_{n,\,max} = 4430$ psi, $f_{v,\,max} = 3430$ psi	
11-14M	$f_{n,\,max} = 31.4$ MPa, $f_{v,\,max} = 25.0$ MPa	
11-16C	2860 psi	
11-16M	22.5 MPa	
11-18C	3390 psi	
11-18M	22.2 MPa	

ANSWERS TO EVEN-NUMBERED PROBLEMS

11-20C $f_{n,\,max} = 30{,}700$ psi, $f_{v,\,max} = 20{,}700$ psi
11-20M $f_{n,\,max} = 60.1$ MPa, $f_{v,\,max} = 32.4$ MPa

chapter 12

12-2C   86
12-2M   *86*
12-4C   75.7
12-4M   *75.7*
12-6C   23,400 lb
12-6M   *81.6 kN*
12-8C   531 lb
12-8M   *251 kg*
12-10C 1260 lb
12-10M *592 kg*
12-12C 103,000 lb
12-12M *38.8 Mg*
12-14C 102,000 lb
12-14M *44.8 Mg*
12-16C W12X72
12-16M *300 × 300 − 106 kg/m*
12-18C 0.404 in. dia.
12-18M *10.8 mm dia.*

12-20C 0.281 in. × 0.562 in.
12-20M *7.27 mm × 14.5 mm*
12-22C $\frac{1}{4}$ in. sq.
12-22M *6 mm sq.*
12-24C W10X49
12-24M *250 × 250 − 64.4 kg/m*

INDEX ————————————————